A BREEDER'S GUIDE TO GENETICS

Relax, It's Not Rocket Science

By

Ingrid Wood and Denise Como

© 2004 by Ingrid Wood and Denise Como. All rights reserved.

No part of this book may be reproduced, stored in a retrieval system, or transmitted by any means, electronic, mechanical, photocopying, recording, or otherwise, without written permission from the author.

The authors shall not be responsible for any injuries, damage, or liability incurred by others in regard to information found in this publication.

ISBN: 1-4140-2478-9 (e-book)
ISBN: 1-4140-2477-0 (Paperback)
ISBN: 1-4140-2476-2 (Dust Jacket)

Library of Congress Control Number: 2003098525

This book is printed on acid free paper.

Printed in the United States of America
Bloomington, IN

Illustrations by Maryann Conran

Cover photograph by Alice Brown, Forever Precious Alpacas
Back cover photograph by Barbara Ewing, Kirov Borzois

1stBooks - rev.02/24/04

Dedication

This book is dedicated to the late Rosa Bernard of Hunolstein, Germany. Ingrid always wanted to grow up to be just like Rosa. Her siblings tell her that she did.

"The fact is that the mortal … is the runner in the relay race where DNA is the baton."

Mahlon B. Hoagland,
The Roots of Life

Acknowledgments

Imagine a professional football player who, although tops in his field, has to write a lengthy proposal each year explaining why he should be permitted to play and get paid for it. Sometimes permission is granted, sometimes it is denied.

Preposterous, right?

The author of a magazine article once stated that scientists do not work hard enough to make scientific concepts and discoveries understandable to the public. At the time, we agreed. It has occurred to us, however, that the vast majority of researchers are probably too darn busy begging and scrounging for grant money to spend any time on such pursuits. After all, some time must be devoted to the actual research! It is sad that our society lavishes millions on entertainers, yet leaves crumbs for the people who enhance and save our lives and those of our animals.

During the research phase of this book, we made personal contact with and sought help and information from several scientists. Starting with Erica Chevalier-Larsen, a young graduate student (whose Senior Honors thesis discussed "A Screen for Genes that Interact with the Dynein Light Chain"), to D. Phillip Sponenberg D.V.M., Ph.D., a professor of pathology and genetics and a world-known authority on mammalian color inheritance, all shared their knowledge without hesitation. They did so with unfailing good cheer, patience, and a refreshingly down-to-earth willingness to explain concepts at our level of understanding. We were never given the feeling that our questions were bothersome or beneath their level of expertise.

Patricia A. Craven Ph.D., a research scientist and alpaca breeder, responded to many e-mailed questions quickly and in detail. Princeton University professor Shirley M. Tilghman not only cheerfully answered questions on genomic imprinting over the telephone, she also immediately posted additional material. Ilene Cottler MD, Ph.D., a busy physician and dog breeder (Triasic English Springer Spaniels), readily agreed to read our manuscript and act in an editorial capacity. She spent several hours discussing the manuscript. Her candid input, both a written critique as well as verbal commentary, was welcome and meaningful.

Nina R. Beyer, V.M.D., a dedicated veterinarian and fellow sighthound fancier, contributed information and valuable assistance to our research. We immediately accepted her generous offer to scrutinize the manuscript. Dale Graham Ph.D., a llama owner and breeder since 1984, carefully read the chapters covering camelid color genetics. Graham, who has her Ph.D. in molecular biogenetics (that's DNA and RNA stuff to us non-scientists), sent detailed comments on what we had written. Knowing what an incredibly busy lady she is, we were immensely grateful to receive such a thoughtful response. Barbara E. Ewing (Kirov Borzoi) helped to present information on the inheritance of color genes in dogs more clearly.

Writing a reference book like this is not easy. A single incorrect or omitted word can radically alter the meaning of a sentence. A **Bb** instead of a **bb** would totally confuse the reader by giving him or her the wrong information. We are both concerned that, through carelessness or ignorance on our part, such "mutations" may appear in this text.

To our great relief, we had the benefit of our dedicated editorial readers. Words cannot express the deep appreciation we feel toward those who provided scientific insight. In addition, Linda McDonald (Sevenoaks Whip-pets) read the entire unfinished manuscript and made many corrections and valuable suggestions. Any mistakes still found in our work are of course our responsibility.

Finally, we must thank our chronically less-than-patient spouses. Their plaintive cries of "Aren't you done yet?" spurred us on (well, sort of). Of course, we can't very well complain about the fact that they missed our company.

We also thank Maryann Conran for the wonderful illustrations she created specifically for this book, and the talented photographers for allowing us to use their photos.

To all who have so generously shared their knowledge about Whippets and alpacas, horses, sheep, goats, rabbits, Borzois, Weimaraners, llamas, Afghan Hounds, Greyhounds — oh, the list is endless! — we salute you!

Thank you all so very much!

Contents

DEDICATION iii
ACKNOWLEDGMENTS v

FOREWORD - An Introduction ix
CHAPTER ONE - Alphabet Soup 2
CHAPTER TWO - The Experts Disagree (With Each Other) 10
CHAPTER THREE - The Cell 14
CHAPTER FOUR - Tinkers to Evers to Chance, or DNA to RNA
   to Protein 24
CHAPTER FIVE - Snips, Snails, and Puppy Dog Tails 32
CHAPTER SIX - Genomic Imprinting 40
CHAPTER SEVEN - Dominant Genes 48
CHAPTER EIGHT - Genetic Probabilities 56
CHAPTER NINE - Lethal Dominants 62
CHAPTER TEN - Incomplete Dominance 66
CHAPTER ELEVEN - Incomplete Penetrance 70
CHAPTER TWELVE - Recessive Genes 76
CHAPTER THIRTEEN - Polygenic Inheritance 82
CHAPTER FOURTEEN - Fact Or Fiction? 88
CHAPTER FIFTEEN - Opposites Attract — Sometimes 96
CHAPTER SIXTEEN - Inbreeding 102
CHAPTER SEVENTEEN - Linebreeding 112
CHAPTER EIGHTEEN - The Co-efficient of Inbreeding 120
CHAPTER NINETEEN - Outbreeding 126
CHAPTER TWENTY - Crossbreeding 130
CHAPTER TWENTY-ONE - Hybrid Vigor 142

CHAPTER TWENTY-TWO - Genetic Anomalies 148
CHAPTER TWENTY-THREE - DNA Testing 154
CHAPTER TWENTY-FOUR - Mutations 160
CHAPTER TWENTY-FIVE - Breeding for Extremes 166
CHAPTER TWENTY-SIX - Fads or Function? 170
CHAPTER TWENTY-SEVEN - Pedigree Power —Pedigree Perils 180
CHAPTER TWENTY-EIGHT - Nurture Helping Nature 188
CHAPTER TWENTY-NINE - An Introduction to the Inheritance
    of Color Genes 196
CHAPTER THIRTY - Color Loci 208
CHAPTER THIRTY-ONE - Color-Coding Rainy 252
CHAPTER THIRTY-TWO - Blue-Eyed Girl 258
CHAPTER THIRTY-THREE - Compromises 268
CHAPTER THIRTY-FOUR - Suri I and Suri II 274
CHAPTER THIRTY-FIVE - Advice (Short and to the Point)
    and Apple Cake 290
POSTSCRIPT — *"A Breeder's Storyquilt"* 294
GLOSSARY 299
REFERENCES 301
INGRID'S LETTER TO OUR READERS 305
INDEX 307
ABOUT THE AUTHORS 313

Foreword

An Introduction

"Genetics: the branch of biology dealing with the phenomena of heredity and the laws governing it."
D.C. Blood & Virginia P. Studdert
Baillière's Comprehensive Veterinary Dictionary, London, Philadelphia, etc.: Bailliére Tindall, 1988

Maybe it is not so surprising that Johann Gregor Mendel (1822-1884), commonly acknowledged as the "Father of Genetics," was an Augustinian monk. Had Mendel been married and the father of a large family, as was customary in those days, he undoubtably would not have enjoyed the long hours and undisturbed peace in which to pursue his studies. More than a century ago, leisure activities were reserved for the wealthy. The vast majority of the population — uneducated and poor — worked backbreaking hours just to provide life's bare necessities.

With rare exceptions, women of the nineteenth century were considered intellectually incapable of learning and retaining scholarly knowledge. They were certainly not expected to make brilliant discoveries in the field of science. If there lived a "Mother of Genetics" during Mendel's lifetime, she toiled in anonymity.

The biologist Robert Pollack would probably bestow such an exalted title on American scientist Barbara McClintock (1902-1992). In *Signs of Life* (1994), Pollack tells us that McClintock "is perhaps the most original geneticist since Mendel and certainly the most self-disciplined person I have ever known." Like Mendel, McClintock met with initial indifference when she introduced the revolutionary idea of "jumping genes" (*transposons*) to the scientific community. Unlike the teaching monk, she gained recognition and admiration for this important discovery before her death.

Mendel himself remained a virtual unknown during his lifetime. Born in Austria, the now world-famous botanist conducted his experiments in a Czechoslovakian monastery, in the town of Brünn (now called Brno, located in the present day Czech Republic). His published discoveries were politely ignored for many years. Only after his death did other scientists recognize the significance of Mendel's work. They used his findings as the foundation for their own forays into the field of genetics.

The pace of genetic discoveries since then has accelerated to the point of leaving the average person bewildered, overwhelmed, and maybe even apprehensive about the subject. The sheer abundance and complexity of information thrust at the public by the scientific community creates confusion and misunderstanding.

The basic genetic knowledge that every breeder of animals should be comfortable with does not come close in degree of difficulty and scope to recent genetic research. Breeders often feel reluctant to try to understand a subject that should be close to the hearts of anyone involved in the process of creating new life — be it in a kennel, a cattery, or a barn. We often find indifference instead of passionate interest and curiosity, and breeding decisions based on beliefs that were disproved years ago.

When Denise Como and I first casually discussed the possibility of collaborating on this book, neither of us felt any sense of urgency to begin the project. Denise, the author of the popular sighthound performance handbook *"So, You Want to Run Your Sighthound"* (now in its sixth printing), was completing her new book *"Sighthounds Afield — The Complete Guide to Sighthound Breeds and Amateur Performance Events."* She has written for *The AKC Afield*, *The AKC Gazette*, *A Breed Apart Greyhound Magazine*, *Celebrating Greyhounds*, the German publication *Der Windhund Freund*, *The Chart Polski Newsletter*, the new magazine *To The Line*, and others. I had just completed the first draft of a book I wrote in German, the language of my country of birth and childhood. Contributions to several camelid publications kept me busy as well.

A few sentences posted on the Internet's Alpaca Site by Camille Mayersky, a llama and alpaca breeder from Oregon, galvanized me into action. After a spirited — and occasionally confrontational — exchange of opinions concerning a particular genetic aspect of breeding alpacas, Camille gently chided the main participants of the heated debate. "Most of the posts were over my head," she complained. "I don't have any background in science."

Camille's comments grabbed my attention. When I phoned her later in the day, I found myself speaking to a warm and articulate woman who obviously cared deeply about her animals. She has been quite successful breeding llamas and alpacas, but expressed an interest in learning more about genetics. Camille did not know where to look to begin her independent study program.

Again, I asked myself why perfectly intelligent people feel such reluctance to avail themselves of knowledge that could possibly prevent costly mistakes and take part of the "luck factor" out of the breeding equation. Public library shelves carry an abundance of books about genetics, molecular biology, and about breeding various pets and livestock — a most diverse and interesting selection.

I concluded that many of these carefully crafted and edited volumes have one problem in common: they are simply not written with the raw novice in mind. Most books on genetics require more than a basic understanding of biology, chemistry, and mathematics. Chapters on mammalian color genetics especially look like so much alphabet soup to a beginner. Authors rarely offer enough practical examples for the reader who is totally unfamiliar with the concepts to reach a deeper understanding of the material.

"You know," one such breeder confessed to me, "between my job, my family, and doing barn chores, I have very little time to read. I am so tired at night that I can't concentrate on all this information. It's too hard to understand."

A Breeder's Guide to Genetics offers basic information to the "scientifically challenged" and to beginners. Our modest book is not geared toward experienced breeders who already possess a sound working knowledge of genetics. We suggest that such veteran readers study the previously mentioned mountain of texts written on the subject by authors with vastly superior knowledge and insight.

The study of genetics, at any level, should not be a daunting and dreaded chore. It should be fun and, especially at the basic level, give one the confidence to delve deeper and unravel the fascinating mysteries of inherited traits. As with any complex issue, it does not produce "instant experts."

Merely memorizing the nomenclature of basic genetic principles does not prepare one to immediately defend strong opinions on a subject as multifaceted as this. Studying them, however, should serve novice breeders and those who decide to approach their breeding program with careful planning to lend it structure and purpose. From our own background of interests and experiences, Denise and I will explain concepts primarily using data about various dog breeds and camelids. <u>We hope the reader will understand that basic genetic principles apply to all species, including Man.</u>

We both believe that no story is too foolish to serve as a simplified example to promote the understanding of scientific concepts. Therefore, I have not shied away from employing whimsical — and sometimes downright silly — descriptions to illustrate necessary points. The more serious-minded scholars among you might find my methods too "unscientific" for your sensibilities. It's okay. We aren't easily offended.

Most breeders, if they are anything like me, do not retain technical information the first time it is presented. Consequently, you will see many little reminders and memory boosters along the way. The few geniuses with perfect memory will just have to suffer through the repetitions.

If, after a quick glance, you decide the chapters on the cell, DNA, and genomic imprinting are too difficult and confusing, just skim over them. Then return to them later to gain a deeper understanding of genetic inheritance. Perhaps that's putting the cart before the horse. So what? That sequence worked well for me, although it is a longer route to take. I am no wizard in the field of science.

My spirits were boosted considerably when I learned that Gregor Mendel failed his admittance exams to study physics and natural science at the University of Vienna — twice! The first time he failed because he had not mastered the material, the second time because he argued with the examiner. You see — there is always hope!

Years ago, I didn't realize that the knowledge of pure science (genetics) is a prerequisite for the <u>true</u> comprehension of breeding programs. My desire to understand the "system" behind the various breeding options available probably made all the difference. I finally overcame my fear of pure science and succeeded where previously I had only experienced confusion and defeat.

When I learned of the enormous contributions German and German-American scientists made and continue to make in the field of genetics, I couldn't help but wonder what contributed to my own genetic make-up. My family did not show much interest in the subject. I do remember my father's reply when my petite sister expressed concern about the size of her truly tiny daughter. "What did you expect?" he asked, looking from my sister to her equally short husband. "Mice make mice."

True enough in most cases, but there are times when, metaphorically speaking, they produce elephants.

I occasionally use human examples to discuss or clarify concepts about animals. We like to think of ourselves as unique and so different from all mammals lacking our speech patterns and opposable thumbs. We are not! The basic components that make up the DNA of dogs, llamas, and all other critters make up the DNA of their two-legged caretakers as well. What is the difference? The sequence in which the nucleotides' bases appear in the DNA chain differs in the various species. Is that it? *That's it!* What's more, specific human DNA sections are identical to those of "other" animals.

The April 2000 *Newsweek* issue featured a short article about gene sequencing in the fruit fly. What is the significance of having the fruit fly's genome "read" and understood? In Thomas Hayden's words: "... because some two-thirds of genes associated with illness in people are also found in Drosophila. The new data should help researchers to further explore the causes of human genetic disease."

Canine DNA research has advanced more quickly due to the interest in genetic defects "shared" by Man and his most faithful companions.

We can now appreciate that gaining genetic knowledge about one species can help us to become more successful breeders of other species.

When I researched data for this book, I was referred to a sheep breeder to learn more details about Spiderleg syndrome. After we had talked for a while, the breeder's wife commented, "You breed alpacas and dogs — why do you want to know anything about sheep?" Some dog breeders subscribe to the *AKC Gazette* but only show interest in the articles about their own breed. Don't be like those people! Read everything!

My two biggest "lightbulb moments" during my ongoing study of genetics occurred while reading books about horse color inheritance and rare livestock breeds.

There is an enormous amount of valuable information available to the breeder who is thirsty for knowledge — it pays to be open-minded and use all the resources you can. View and use our guide the way it is intended — as a smorgasbord of information, and as your first "baby steps" toward understanding the basic knowledge needed to follow the more complex work of other authors. Nothing will give Denise and me more pleasure than hearing that, after reading *A Breeder's Guide to Genetics*, you felt inspired to tackle the reference books we've used and listed.

The opinions expressed in this book are, of course, solely those of the authors. In other words, do not go chasing after our editorial readers if you don't agree with us! Those generous and hardworking people made corrections and offered suggestions — but they did not dictate editorial policy to us. If you do not agree with various passages, please feel free to share your comments with us.

Why did I ask Denise to join me in this project? Besides admiring her lively intellect and enjoying her (at times) wicked sense of humor, I also desperately needed her considerable editorial skills and knowledge of publishing to get this project moving. Then there was the little matter of putting the manuscript together. I am, for all practical purposes, quite computer illiterate. When I return home from my job, and I've taken care of household chores, cleaned the pastures, fed and watered the alpacas and the whippets — well, I am simply too tired to learn "computer."

All this being said, let us begin our journey into the world of genetics.

Chapter One

ALPHABET SOUP

CHAPTER ONE

Alphabet Soup

"Genetic nomenclature is subject to a number of conventions, and these are variable in much of the literature."
D. Phillip Sponenberg D.V.M., Ph.D., *Equine Color Genetics*, 1995

In order for participants to communicate intelligently, all specialties develop their own nomenclature, or names for things. A novice must learn this "language," or important data remains inaccessible. You may be initially confused by the fact that phraseology may vary according to the author. For example, some geneticists interpret the term *crossbreeding* as meaning either crossing animals between two separate breeds (as in breeding a Saluki to a Borzoi), or staying within a breed but crossing to another breeder's line (as in breeding my alpaca to your alpaca, who have come from totally different ancestors). In *A Conservation Breeding Handbook* (1995), D. Phillip Sponenberg D.V.M., Ph.D., distinguishes between the two by referring to the latter as *linecrossing*.

As a fledgling alpaca breeder, I was quite confused to see llamas, guanacos, alpacas, and vicuñas listed as separate species. I always understood that members of one species cannot produce offspring with those of others. In the rare cases that they do, such as horse and donkey, the offspring is sterile. Individuals must be genetically compatible (identical chromosome count and sufficient genetic similarity) for the female of a species to conceive, never mind give birth to, viable fertile offspring. A human and a chimpanzee cannot have children together, though they share close to 99 percent of their genetic make-up — and thus are genetically closer to each other than, for example, a chimpanzee and an orangutan.

Professor Robert C. King gives this definition of *species*: "One or more populations, the individuals of which can interbreed, but which in nature cannot exchange genes with members of other species" (*A Dictionary of Genetics*, 1978).

This is not so with the four camelids mentioned above. They can and do crossbreed with each other and produce fertile offspring. I was therefore happy to see Eric Hoffman's succinct comment in the *Alpaca Registry Journal* (Vol.V, No.1, Spring 2000): "They are officially listed as separate species, but by most scientific definitions, they could be considered different breeds instead."

Dr. Rigoberto Calle Escobar, professor of sheep husbandry in Peru, wrote in *Animal Breeding and Production of American Camelids* (1984): "Secondly, I must refer to the fact that Llamas and Alpacas are not definitively differentiated from a taxonomic point of view, and it seems that they correspond to breeds rather than to species."

Llama owners enjoy an amazing array of activities with their animals. Called "the ships of the Andes," North American llamas carry their owners' food and other provisions on hiking trips, pull carts, and are welcome visitors in schools and nursing homes (photo by Morning Star Ranch).

A BREEDER'S GUIDE TO GENETICS
Relax, It's Not Rocket Science

I would therefore classify the two different types of alpacas, Huacaya and Suri, as breed *varieties* (<u>not</u> breeds as they are defined by ALSA, the show organization), similar to the coat varieties found in several breeds of dogs.

Llamas and alpacas have been crossbred (llamas bred to alpacas) to various degrees for thousands of years by South American pastoralists. This practice became more prevalent after the Spanish virtually destroyed the more sophisticated breeding programs adopted by the Incas. In *Lamas — Haltung und Zucht von Neuwelt-Kameliden* (8.Jahrgang, Heft 2), Dr. Martina Gerken reports on a presentation made at the Second World Congress for camelids held in Cusco, Peru, in November 1999. A genetic DNA analysis of more than one thousand live and mummified camelids identified roughly 90 percent of all <u>tested</u> alpacas as being crossbred with llamas. For llamas, the number of blood samples showing alpaca genes was 40 percent. Dr. Jane Wheeler's discovery showed that hybridization "was a far greater problem than anyone had suspected" (Heather Pringle, *Discover Magazine*, April 2001).

Geneticists Dr. Miranda Kadwell and Dr. Michael Bruford confirmed with their research at the Institute of Zoology in London, England, what Dr. Wheeler had long suspected after working with camelid skeletons: "The vicuña are the most likely ancestor of the alpaca, and the guanaco are the most likely ancestor of the llama" (Bruford, *Discover Magazine*, 2001).

Eric Hoffman reports in *The Complete Alpaca Book* (2003) that "Wheeler's work has resulted in reclassification of the alpaca from *Lama pacos* to *Vicugna pacos*."

What about wolves, coyotes, and dogs? They are all classified as members of the *genus* Canis and considered separate species, yet they produce fertile offspring when interbred. It's all confusing to a lay person (a non-specialist).

We can, of course, adopt the stance that Colin Tudge assigns to post-Darwinian biologists: "No definition of species can be perfect, so it is foolish to try and frame one" (*The Engineer in the Garden*, 1993). He believes that the "viable offspring" scenario works well enough. His detailed observations on the subject are quite interesting.

Genetic engineering makes the exchange of genetic material between species possible. The genomes of Scottish Blackface sheep were "engineered" by Ian Wilmut and Keith Campbell (creators of Dolly, the cloned sheep) to carry the human gene for the enzyme AAT. One day soon, those who suffer from Cystic Fibrosis will get relief due to the contributions made by these *transgenic* animals. Such genetic action is impossible to achieve with good old-fashioned sex.

What about the term *breed*? Dr. King defines *breed* as "an artificial mating group derived from a common ancestor for genetic study and domestication." Dr. Sponenberg quotes Juliet Chilton-Brock: "... a breed is a group of animals selected to have a uniform appearance that distinguishes them from other groups of animals within the same species."

The glossary of one of Dr. Sponenberg's books gives this definition: "Breed - a group of animals that are similar enough to be logically grouped together, are distinct from others of the same species, and when mated together will reproduce this distinguishing type." For example, a Borzoi bred to a Borzoi produces a litter of puppies that all exhibit the typical Borzoi phenotype (the Borzoi "look"). We call this "breeding true."

Since all authors give further details and include glossaries with their work, you will soon catch on to the exact meaning of their chosen terminology or vocabulary. In conversations and correspondence with others, you might have to specify your interpretation of a definition to avoid misunderstandings. Most terminology is standard, however, so the few exceptions should not pose a major problem. If you are still not sure of the intended definition, do not be afraid to ask!

Genes can be *dominant* or *recessive*. In the next chapter, you will learn in detail how a single dominant gene expresses itself (such as the **A** in the **Aa** combination). On the other hand, it takes both copies of a recessive gene (as in **aa**) for the trait they "code" for to surface in the animal's phenotype. You will be relieved to discover that the experts usually assign an uppercase (capital) letter to the *dominant* gene and a lowercase (small) letter to the *recessive* gene. This knowledge will serve you well when you study color genetics.

If you read, for example, that a Labrador Retriever is also **BB** or **Bb** in addition to carrying the dominant **A** and **E**^{m}, you know that the dog's coat color is black. If it is described as **bb**, you know that the dog is chocolate or liver (brown) colored — this will make more sense after you read the chapter on color inheritance.

You will learn that the words *gene* and *allele* may be used interchangeably where <u>alternate forms</u> of a gene exist in a population — for example both **B** and **b** or both **A** and **a**.

One scientist will show alleles as I just explained. Another may show the combination **aa,** coding for black fur in cats, as A^aA^a. This little example, by the way, helps me to make the point that dominant traits in one species or even in one breed might be recessive in others.

Another goodie in the bag of scientists' tricks is the use of dashes behind dominant alleles when the second allele does not contribute to an animal's phenotype. The **BB** and **Bb** combinations can also be shown as **B-**. Since both **BB** and **Bb** result in a black coat rather than a liver coat, the second allele is immaterial (its importance changes, of course, when the dog is bred).

Some authors will use **B+** to identify the dominant and a plain **B** to list the recessive allele. Keep stirring that alphabet soup!

Fanciers of various breeds that belong to the same species often use different names for the same color. *Wheaten* (a tannish color) in a terrier or Irish Wolfhound might be called *deadgrass* in a Chesapeake Bay Retriever, *fawn* in a Greyhound or Whippet, *cream* in a Saluki, *apricot* in a Borzoi.

Leaping from one species to another, while retaining all or part of the nomenclature of the first, can also lead to confusion. Black and white horses were traditionally called piebalds, while horses of other colors with white spotting were called skewbalds (these terms have fallen into disfavor with horse owners because they lump the various spotting patterns together). I've heard owners describe alpacas of any color and showing any spotting pattern as piebalds, thereby confounding horse owners who are still familiar with these outdated terms.

Some scientists spell the scientific name for red pigment *phaeomelanin*. Others choose *pheomelanin*. I would be happy to use either spelling — if the experts could make up their minds!

Lucky are those breeders whose registries demand blood-typing for the purposes of parental identification, but even registries do not always tell the complete genetic story. Our alpaca Kalita is correctly registered as a white Huacaya female. To the eye, her fleece is snowy white, but genetically speaking Kalita is either black or red (brown). An entire series of letters informs the geneticist about an animal's color genotype as well as its phenotype. Don't get excited! You will be able to "read" them by the time you've finished this book. I promise!

Later, we will explore how genetic research on coat- or fleece color is greatly hampered by breeders reporting the wrong color on registration certificates or by being mistaken in the true identity of a sire. In that case, we don't just have alphabet soup, we have an entire soup kitchen!

Don't think breeders are the only ones who disagree at times. The experts we depend on for information, advice, and guidance frequently engage in their own battles.

Chapter Two

THE EXPERTS DISAGREE (WITH EACH OTHER)

CHAPTER TWO

The Experts Disagree (With Each Other)

"Not everyone agrees with all the conclusions."
Top Science and Health News (Reuters), discussing research on the human genome, February 11, 2001.

In a strange way the study of genetics appeals to, as well as annoys, those with mathematically inclined minds. Precise formulas, used to determine the probabilities of inheriting a specific trait, and the co-efficient of inbreeding appeal to those with "left-side" brain function. Certain genes, as we will discover, are inherited in a very orderly, easily determined fashion. They can be "pulled" out of the sire or dam like sweaters out of a superbly organized wardrobe. There are no surprises — good or bad — merely the practical application of a well-researched and time-tested genetic principle. Reported deviations often turn out to be mistakes made by a breeder or the person recording the data.

Unfortunately, only a small section of the genetic "wardrobe" is so orderly and predictable. You can reach into your closet expecting to find a long black skirt, and end up with a short plaid one instead. You shake your head. How can this be? Every other time you reached into the closet, the result was what you expected.

These surprises often lead breeders to tell potential buyers that "breeding is all just a crapshoot anyway." Well, it really isn't. Besides spontaneous mutations (which are rare), there is "rhyme and reason" to the results of our breeding decisions. We might not always have the knowledge to understand how they came about. Well-educated and superbly trained geneticists do not always know either. You should realize that the experts often disagree with each other, or at they least question the conclusions of their peers. This can upset and confuse new bright-eyed student of genetics.

You may feel you have finally grasped and digested the information on German Shepherd Dogs in Clarence C. Little's book, *The Inheritance of Coat Color in Dogs* (1957) — until you read Malcolm B. Willis's *Genetics of the Dog* (1989). Confidently opening what many breeders consider the "bible" of dog breeding, you reach page 89, smugly expecting the identical information. You frown in dismay when you discover that Dr. Willis, referring to a book he wrote in 1976, disputes some of Little's premises about the *Agouti series*. Another fellow named E. A. Carver is thrown into the picture, further rejecting two of Little's theories. By the time you read Leon F. Whitney's opinions about recessive blacks in that breed, which clashes with data provided by Carver, you are ready to scream in sheer frustration. At that point, it occurs to you that your new litter of puppies (who don't care about genetics) could energetically shred all the reference books into little pieces and you wouldn't mind in the least!

How can we, as lay people, be expected to comprehend the issues with so much conflicting information in print?

I cannot offer a ready-made solution to this little (no pun intended) dilemma. The best advice I can give you is to read as many authors as possible. Question experienced breeders about their own observations. Pick their brains for as long as they will allow. Visit kennels and farms and employ your own powers of observation. Ask about the relationships between the individual animals you see. Trust me, armed with only the incongruent material you've read, you can still arrive at conclusions and make informed decisions.

People who have the patience to solve the *New York Times* crossword puzzle, as well as passionate readers of mystery novels (following and dissecting clues with the excited snuffle of a bloodhound tracking a criminal), will do very well in this pursuit. I suggest leaving the sub-topic of coat- or fleece color inheritance until you are comfortable with the basics. That's probably disappointing advice to the owners of llamas and alpacas, whose animals come in a virtual rainbow of colors — but it's only sensible and will save you from suffering much frustration. After all, you wouldn't attempt to study algebra until you've mastered basic computation skills (unless you are Albert Einstein). As you'll discover, there's an exception to everything in this branch of science called genetics.

There were times that, while doing research for this book, I felt sorely irritated when the "facts" presented by one author conflicted with those of other authors or experts. For example, one author described Tortoise-shell and Calico cats as genetically and phenotypically the same. Not knowing much about cats, I took his word for it. Only Denise's insistence that further research was needed — combined with information given by a friendly alpaca breeder who owns cats — set me straight. A calico cat is a tortoiseshell with large white markings. In another case, I probably read four different estimates for the number of carriers of Cystic Fibrosis in the Caucasian population.

One author stated that "... we would not think of selecting for color characteristics in an albino line" (Srb, et al.). At the time, that statement made perfect sense to me. Then I learned from the famous mouse fancier W. MacKintosh Kerr that albino mice, who often genetically proved to be chocolates or champagnes, were indeed used to establish and improve specific colors in certain strains. Kerr used a color-tested albino cross as a foundation animal in his successful attempt to "manufacture" (Kerr) a superior line of fawn-colored mice.

Researchers at one time listed black, red, and chocolate as the three basic mammalian pigments. We now know that chocolate/liver are biochemically the same as black. "They are," Dr. Sponenberg firmly assured me, "just packaged differently." Modern authors describe black and red as the _two_ basic pigments carried by mammals.

I remind myself — and remind you — that genetics is not a static science. We must appreciate the constant flow of new information. Gifted scientists make new discoveries precisely because they refuse to accept "truth" as they know it. They discard old notions that are proven incorrect and embrace newly found knowledge and facts.

Those of us with less probing minds must strive to keep them open and nimble — or be hopelessly left behind.

You may not understand everything the first time you read it. Although Denise and I made every effort to explain information in detail, there comes a time when you can't simplify things any further.

Learning and truly understanding even the most basic facts takes some dedication. The study of genetics is too vast an area for even scientists to be familiar with all of its many facets.

As I've suggested before: skim over the next three chapters, but promise me that you'll return to them later. Their study will clarify the contents of all other chapters enormously. To those of you made of sterner stuff, proceed to Chapter Three.

Chapter Three

THE CELL

CHAPTER THREE

The Cell

"Cell: the basic structural unit of living organisms."
Blood & Studdert, *Baillière's Comprehensive Veterinary Dictionary*, 1988

The study of genetics is essentially the study of the cell, a tiny protoplasmic body capable of independent reproduction. The cell has revealed its secrets somewhat grudgingly over the centuries. Our knowledge of its contents and their functions has nevertheless grown enormously during the last few decades.

When Gregor Mendel conducted his experiments, documented his findings, and reached his conclusions, he only had a vague idea of how the actual process of inheritance worked. Any concrete, scientific validation given to Mendel's theories had to wait until more sophisticated tools were available to the pioneers of "modern" genetic research. Even after microscopes became more effective, our understanding of cell structure and genes remained superficial for a long time.

Mendel intuitively recognized that a component other than blood must be responsible for passing on physical traits. Until then, people believed that the blood of the parents mixed or blended together in their children, thus creating the millions of unique individuals inhabiting the earth. Unfortunately, misleading expressions like *pure blood* and *blood relatives* are still in use today.

Increasingly powerful microscopes and newly invented dyes that revealed previously undetected cell details moved the study of genetics forward.

Several years ago, my friend Barbara requested my assistance during an artificial insemination procedure she performed on her Borzoi bitch in the comfort of her home. My friend urged me to look at a small semen sample she placed under a microscope. Even though it wasn't state of the art equipment, I was completely amazed at the clarity and vivid details of the swarming school of sperm cells, all literally swimming for their lives.

Only two decades after Mendel's work, the German biologist August Weismann (1834-1914) advanced the theory that the body produces two different types of cells. It does! *Germ cells* (egg and sperm) are manufactured by the reproductive organs to insure the continuation of species. S*omatic cells* concentrate on supporting germ cells in this quest by organizing the growth of tissues and organs. It's the germ cells in animals and humans that pass on genetic material to the offspring.

A trio of German scientists, Mathias Schleiden (1804-1881), Theodor Schwann (1810-1882), and Rudolf Virchow (1821-1902) contributed to the revolutionary cell theory that recognized the following important concepts:

- Any living organism is composed of one cell or multiple cells.
- All chemical activity governing life happens within these cells.
- New cells can only be created from already existing cells.

Any grade school science book depicts a simplified drawing of a cell. We have the *membrane* (picture the skin of a peach), the *cytoplasm* (the flesh of the peach), and the *nucleus* (the peach pit). The membrane allows nutrients to enter the cell. The cytoplasm is like a little chemical factory, making protein and breaking down sugar for energy, among other critical activities. The nucleus, the inner core of the cell, is covered with a thin, porous membrane, allowing chemicals to enter it from the cytoplasm. It contains most of the genetic material and is the control center of major genetic action in the cell.

A structure called the *mitochondrion* is located in the cell's cytoplasm. In *Gene Future* (1993), Thomas F. Lee tells us that the mitochondria found in it "are vital to the cell, for it is there that oxygen is consumed to complete the breakdown of sugars and acids and the resulting energy is captured for the cell's use." Mitochondria carry comparatively little genetic material (DNA) which is only inherited maternally (from the dam). The sire's sperm do not contribute *mitochondrial DNA* to the offspring. Mature sperm combine only their nucleus with the egg during fertilization.

A BREEDER'S GUIDE TO GENETICS
Relax, It's Not Rocket Science

How exactly does the creation of new life take place? What happens before we can welcome those squiggly, squirmy puppies, the long-legged foal, the nest of tiny rabbits, and the fluffy alpaca- or llama cria into our world?

The cell nucleus provides the answer. To be more precise, it's the material scientists call *chromatin,* a complex of *Deoxyribonucleic Acid* (DNA) that directs the synthesis of protein. We can think of protein as the building blocks of the cells.

Let's briefly discuss cell division itself, then we can worry about the "before and after." The first cell that starts the long chain of events culminating in birth is formed by the union of two *gametes*: one egg (*ovum*) and one sperm cell (*spermatozoon*). The result of this union is a *zygote*. This original cell does not rest on its laurels, but is driven to feverish activity. It immediately divides itself in two. Each of these cells divides in two, and so on and so on.

Even a mathematically challenged person can visualize how quickly cells multiply exponentially, eventually reaching into the billions. They go on to form an entire body. The division itself does not tell the whole story. Remember the chromatin in the cell's nucleus? Just before a cell divides to promote growth, the structure of the chromosomes condenses, making the chromosomes more visible under a microscope. This process involves the formation of tighter bonds between cell protein and DNA. For a long time, scientists thought each chromosome represented one gene. What kept nagging at them was the fact that there just weren't enough of them to create the unbelievable diversity of complex traits found in many living organisms.

Research and more sophisticated microscopes proved the geneticists to be correct. Animals and people have fewer than one hundred chromosomes in each cell, but each chromosome carries hundreds of genes. All members of a species carry an identically fixed number of chromosomes in each body cell (egg and sperm cells are exceptions, as we will discuss shortly).

Each human cell has 46 chromosomes, a camelid has 74, a dog has 78, a sheep has 54, and each horse cell has 64 chromosomes. Here we have our first example of conflicting information given by the experts. In a workshop handout, Dr. J. Koenig lists a camelid's cell as having 78 chromosomes. Eric Hoffman, Murray E. Fowler, D.V.M., and Professor

Rigoberto Calle Escobar give 74 as the correct number. I've decided to go along with the majority. For our purposes, it really doesn't matter.

Chromosomes and genes are made of *Deoxyribonucleic acid* (*DNA*). DNA is the substance responsible for the *genome* (the total genetic material) of the new baby tottering around in your pasture.

What happens to the chromosomes right before a cell divides?

These chromosomes are busy little beavers at that critical time. Like a class of children dividing themselves into two teams for a kickball game, half of them rush to one side of the cell, while the other half swarm to the other side. As the cell divides, each pair of chromosomes is randomly split between the new daughter cells now forming.
"What's this?" you ask. "Does this woman know what she's talking about? If 74 (llama) chromosomes split up, the daughter cells will only receive 37 each, the next 18.5 — no, this does not make sense!"

You are correct! To prevent them from eventually reaching a countdown to zero, each chromosome builds an exact replicate of itself just before hightailing it to one side of the cell or the other. The still-connected halves of each replicated chromosome are called *chromatids*. Think of them as identical twins holding hands. For a brief moment, before the cell divides, a llama cell contains 148 chromosomes, making it a *tetraploid* cell. Division restores the original number of chromosomes to each cell. This magical cell action is *mitosis*.

What about the division of the cells we label germ cells, which produce *ova* (the plural form of *ovum*) and *spermatozoa* (plural of *spermatozoon*)? *Meiosis* is the somewhat more complicated and lengthy process of halving the chromosomes in a germ cell. Germ cells, once they're ready for action, travel with only half the genetic luggage that somatic cells must carry around — they are called *haploid cells*. When egg and sperm unite during their grand fertili-zation rendezvous, the llama papa brings 37 chromosomes to the table, and the llama momma pitches in with 37 — the llama zygote ends up with 74 once again, forming a *diploid* cell.

You will see different versions of this information in other textbooks. From a practical standpoint, such an account is acceptable. It is, however, not entirely factual in a scientific sense.

Germ cells wait in the animal's body for their moment of glory in a sort of unfinished state. They're not even called egg and sperm at that time, but *oöcytes* and *spermatocytes*. These *primordial germ cells* are all *diploid*, meaning they carry the <u>full chromosome count</u>. As I've mentioned earlier, they become *haploid* during meiosis. The female gamete actually <u>remains</u> diploid, until the sperm's contact with the outer membrane of the female cell causes the oöcyte to develop into a haploid cell. Once fertilization takes place, the new cell is diploid. Interestingly, each primary spermatocyte produces four functional sperm cells. In females, meiosis of an oöcyte results in only one functional egg cell. The other three cells — that do not contain cytoplasm — are called *polar bodies* and are unceremoniously discarded.

Diploid, schmiploid — don't let this drive you crazy! The fact is that a new cell is created, and a brand new individual is developing. Each parent contributes exactly half of the offspring's genome. Chromosomes are contributed randomly, which is why all creatures on earth are truly unique individuals in the true sense of the word. Identical twins and clones like the sheep "Dolly" belong in a category of their own, yet even they have subtle differences due to special genetic and environmental influences. It might surprise you to find out that Dolly, according to her creators, was not a true clone, but a "DNA clone." Dolly's cell nucleus and cytoplasm, along with the latter's mitochondrial DNA, came from two different sheep. Identical twins are genetically closer to each other than Dolly was to the animal she was cloned from, and the one she phenotypically resembled — the nucleus donor.

If the llama baby pronking (also called pronging or stotting) exuberantly at dusk began life as the joining of two germ cells, where did the somatic cells suddenly come from? That's a good question!

Think of the first *diploid* cell as your basic little Mom-and-Pop store selling craft supplies to build everything from toes to hair, eyes to skin, size of the nose to shape of the ears. This one store expands into a small chain of stores. Eventually it becomes too cumbersome for all stores to offer the entire inventory for sale, and management decides to specialize.

Now some stores only sell supplies to manufacture blood, others only bone or brain cells (is this a success story, or what?). A few of the Mom-and-Pop operations remain, just in case there is demand for the entire chain of events to be repeated at a future date.

These primordial cells, ever mindful of their exalted status in the body, do not mix and mingle with the somatic peasants. Shortly after fertilization, they travel to special places and are later called the testicles and ovaries. When the organism is sexually mature and ready to reproduce, the spermatocytes and oöcytes develop into sperm and ovum, respectively — and the cycle begins anew.

Bear in mind that each somatic cell in an organism carries an identical set of genes or genetic "blueprint." Red blood cells are one of the few exceptions.

You may ask how — if all cells carry the entire DNA specific to an individual — the somatic cells can specialize. At crucial times, most genes in those cells are "turned off" by a chemical process. In the bone-making cells, for example, only the genes responsible for making bone stay "turned on." The body's cells are truly amazing little factories.

Let's address the concept of *pairing*. Chromosomes come in *matching pairs*, with the exception of the **XY** chromosomes. You will customarily see, for example, a llama's chromosome count expressed as 37 pairs, rather than 74 singles. The scientifically correct way to describe this is 36 pairs and 2 sex chromosomes. Genes in the *primordial* and *somatic* cells therefore appear in *pairs*, one from the mother (dam) and one from the father (sire). In other chapters, you will read how the expression of a particular trait (certain coat- or fleece color, for example) is determined by one or more *pairs* of genes (alleles).

The term *allele* is used interchangeably, for all practical purposes, with the term *gene*. You may want to think of genes as ice cream and alleles as the different flavors. If you ask for strawberry or vanilla, it is not necessary for you to add the word ice cream — it's understood that you're simply asking for a particular flavor. Alleles are the "flavors" a gene can have.

Let's define it in more detail. The word *allele* is actually a shortened version of *allelomorph*. The glossary of this book tells you that an allele is an *alternate form of a gene*. A good example is the allele for black (**B**) coat color versus the one for liver (**b**) color. An animal can be **BB** (black), **Bb** (black), or **bb** (liver). Both **B** and **b** represent an allele. The combination of both parental alleles determines the *genotype* (the genetic make-up) as well as the *phenotype* (the physical appearance) of the offspring. An animal's genome is the result of such pairings.

	Genotype	Phenotype
BB	black-black	black
Bb	black-liver	black
bb	liver-liver	liver

If you like analogies, think of the body as a genetic Noah's Ark — almost everything comes in pairs.

It is helpful to learn the meaning of several other important terms. The pairing of two identical alleles such as **bb** or **BB** is *homozygous*, while the **Bb** combination is *heterozygous*. Each pair of genes or alleles occupies a specific *locus* (think of a "location" or genetic address) on *homologous* (paired) chromosomes. Since chromosomes come in pairs, each of the two chromosomes can have only one allele at any particular locus. Together, the two alleles determine or contribute to the expression of a *trait*.

"Okay," you say, "a little earlier in the chapter you used ice cream as an example to explain the word allele. Ice cream comes in more than two flavors. What about the others?"

That's a good question. You're actually asking about a concept called *multiple allelism*. Instead of ice cream flavors, we'll use mitten colors to explain it.

You only have two hands, so you can only wear two mittens (alleles) at that particular location (locus) — or for that matter, at any loci.

There are black, red, green, yellow, and purple mittens circulating among the members of your population. You've inherited one red and one green mitten from your parents. You are, however, simply nuts about purple. That's too bad! As an individual, you cannot wear more than two mittens. If you want your children to definitely inherit a purple mitten, your only choice is to find a mate with two purple mittens. This positively ensures that all your children will inherit one purple mitten. The other mitten will come from you, and it will be either green or red. Your children will wear either purple-red or purple-green combinations.

Think alleles instead of mittens. You have now grasped the concept of multiple allelism. I will refer to it again in our discussion on color inheritance.

The *random inheritance* of chromosomes explains the unique genotype and phenotype of all individuals within a species.

"The members of a pair of alleles separate cleanly from each other when an individual forms germ cells" (Srb, Owen & Edgar, *General Genetics*, 1965). During this process, all allelic combinations such as **BB**, **Bb**, or **bb** separate again under what we call the *Principle of Segregation*. The **Bb** (black/liver) combination is therefore not passed on by sire or dam as a unit. One offspring may inherit **B**, another one **b**. With the **BB** or **bb** parents, it obviously does not matter which allele they pass on to their children. The entire genome of your dog, llama, sheep, or alpaca contains possibly tens of thousands of such genes (alleles).

A concept called *Mendel's Law of Independent Assortment* helps us to see the "big picture." The allelic combinations segregate again, and they do so independently of other combinations. Let's assume that both a bitch and a male are **Bb Tt** (two separate loci). By the time the bitch's oöcytes become viable (haploid) gametes, the individual eggs may be **BT**, **Bt**, **bT**, or **bt**. Now picture the male's sperm cells with any one of those four combinations. Take out a piece of paper and figure out the possible number of combinations these eight cells can produce in their offspring! At the end of the calculations, a puppy could carry **BB TT** or **bb tt**, or many other possibilities in between.

This concept remains somewhat nebulous until you envision the physical traits associated with the letters. In canine genetic language, **T** represents **ticking** and **t** represents **no ticking** (for non-dog readers, *ticking* refers to a pattern of small dark spots or "freckles" as seen on some hound breeds). B represents black coat color, b codes for liver. At the **B** and **T** loci in our example, the four eggs and sperm cells may therefore each carry allelic combinations coding for **black/ticking**, **black/no ticking**, **liver/ticking**, and **liver/no ticking**. Now work out these translations for the zygotes. Amazing varieties, aren't they?

Because of the effects of segregation and independent assortment, genetic material is not passed down through the generations in neat, intact packages. The often-used explanation that the offspring inherits exactly 25 percent from each grandparent is misleading in its stark simplicity. While that is theoretically possible, it is highly unlikely. Studying the format of a pedigree unfortunately reinforces this erroneous belief.

A frequent genetic occurrence that scientists call *crossing over* further contributes to genetic diversity. We can think of it as genetic recombination or "shuffling" of genes. During meiosis, while germ cells are being formed and chromosomes are in the chromatid stage, genes originating from the dam's chromosomes often change places with genes at the same locus on the sire's homologous chromosomes. To quote Dr. Malcolm B. Willis: "... alleles carried on a particular chromosome inherited, let us say, from the sire, can become separated from those carried elsewhere on the chromosome and instead become linked up with those which originally stemmed from the dam" (*Genetics of the Dog*, 1989). Genes located close to each other on a chromosome tend to stay together; those being far apart have a greater chance of being separated. It is crucial that breeders understand this.

What exactly is DNA? The news media has certainly been abuzz with articles about DNA research. To understand DNA is to comprehend genes and their role in breeding animals at a deeper level.

Chapter Four

TINKERS TO EVERS TO CHANCE, OR DNA TO RNA TO PROTEIN

CHAPTER FOUR

Tinker to Evers to Chance — or DNA to RNA to Protein

> *"Gene: the unit of heredity most simply defined as a specific segment of DNA, usually in the order of 1000 nucleotides, that specifies a single polypeptide. Many phenotypic characteristics are determined by a single gene, while others are multigenic. Genes are specifically located in linear order along the single DNA molecule that make up each chromosome."*
> Blood & Studdert, *Baillière's Comprehensive Veterinary Dictionary*, 1988

Getting back to my father's little joke: why do mice make mice? Why do llamas make llamas? It's all "in the genes"

What exactly <u>is</u> a *gene*? Now don't get yourself in a tizzy and immediately assume that this chapter will be too difficult to understand. The beauty about how molecular mechanisms in genetics work is the fact that, despite their complex and overwhelmingly complicated details, the basic (okay, <u>very</u> basic!) models can be understood by people like us.

Of course, it helps to have a very knowledgeable and very patient friend like Pat Craven, the owner of Cherry Ridge Alpacas. Pat spent a good chunk of time correcting my mistakes and clarifying "pure" scientific concepts for me. Alpaca breeders should read Dr. Craven's excellent article, "Basic Molecular Mechanisms in Genetics," featured in the Summer 2000 issue of *Alpacas Magazine*.

Years ago, when I first noticed articles about DNA in the media, I initially and very naively believed DNA to be a separate substance from the genes. I thought it somehow <u>led</u> scientists <u>to</u> the gene and their identification of, for example, recessively carried defects. I had completely missed the point that DNA <u>is</u> the genes. This misconception was at least partially caused by scientific nomenclature (how things are named). How many non-scientific people had ever heard of DNA back in the seventies?

If you are routinely asked to pass the salt and pepper (genes) during a meal, you'll be confused when a guest requests the whatchamacallit (DNA). You guess that he's referring to salt and pepper, but you're not sure. (Wait, my analogies get better!) My mission here is to explain the whatchamacallit in such a way that it aids you in making educated and sensible breeding decisions.

Let's begin with a little history. Once chromosomes and genes were identified, scientists clung for decades to the belief that genes were made of protein. This is not surprising — in addition to water, the bodies of mammals are composed mostly of protein. Oswald Avery was the first scientist to prove that DNA, not protein, is responsible for the inheritance of all traits. A tiny wisp of a man, this biochemist nevertheless stands as a giant among his peers. Modest and unassuming, even shy, he was adored by his students. Avery was never awarded the Nobel Prize for his incredible achievement — which is one of life's great injustices. Quite simply, the scientific community was not at all convinced that this tireless, careful worker had proven his point beyond a shadow of a doubt.

In 1950, Martha Chase and her partner Alfred Hershey showed that genes are made of *Deoxyribonucleic Acid*, or *DNA* for short. DNA consists of bases, phosphate, and sugar. The experiments done by Chase and Hershey were apparently more convincing than Avery's in their "elegant" clarity (scientists love that term and use it regularly to describe experiments and research results). Even the biggest skeptics had to capitulate when presented with their results.

That still left scientists in the dark about the exact structure of the "stuff of life." Enter Dr. James Watson, born and raised in Chicago, and Dr. Francis Crick, a British chap. Working together, they conceived the model of the *double helix* and presented it to the scientific community in 1953. (You may want to read Dr. Watson's very amusing book about that time: *Genes, Girls and Gamov*, 2001. The man obviously never heard that gentlemen kiss but don't tell.) Their brilliant imaginations opened the door to research and discoveries that leave me breathless. Such discoveries would have certainly astounded Gregor Mendel. Did you know, by the way, that Mendel's Christian name was Johann? He took the name Gregor when he became a monk.

How can a non-scientist understand these concepts? Begin by picturing DNA in the shape of a flexible ladder. The ladder is much too long to fit into the cell's tiny nucleus for storage — it is therefore twisted and looks somewhat like the tightly coiled shape of the popular rope chew-toys that dog owners give to their pooches to "floss" their teeth. No traveler packs his suitcase fuller and neater than Nature storing DNA in a cell's nucleus.

Segments of DNA are densely packed within a chromosome. Geneticists describe the DNA of higher organisms such as mammals as a *linear molecule*. Dr. Craven, giving more detailed information in the *Alpacas Magazine* article, writes that DNA is composed of "two complementary strands of nucleotides that form a double helical structure." Shortly before a cell divides, the DNA helix unwinds and replicates itself. In the previous chapter I used the term *chromatin* — which is DNA packed together with protein.

Does this genetic copying machine ever make mistakes? Yes, it does — but actually only rarely when you consider the enormous amount of copying that takes place before the zygote has developed into a fully grown llama, dog, or sheep. Just like the owners of any well-managed assembly line, the cells employ a "mechanic" who catches mistakes and repairs them. A few might escape the notice of our busy worker — otherwise known as an *enzyme* — and thus produce mutations. Copying of certain cells actually continues throughout the entire lifespan of an organism.

You may also think of the enzyme as an editor correcting a manuscript. Shortly, you'll understand why.

"All right," you say, "so far, so good. All this information still doesn't exactly clarify what a gene is and how it works."

Genes are identified as <u>specific discrete portions of the DNA molecule that are arranged along a chromosome in stable, linear sequence</u>. Translation: a gene is a specific section of a long string called DNA. The word *sequence* is the key to understanding this concept. How so?

Chemically speaking, DNA is composed of "building blocks" consisting of a series of four related *nucleotides* (chemical units). Each nucleotide contains a sugar (deoxyribose), one or more phosphate groups, and one of four different bases. Scientists talk about a chain or *sequence* of bases, giving them the lyrical names of *adenine* (A), *thymine* (T), *cytosine* (C), and *guanine* (G).

The authors of a book mentioned at the end of this chapter give an example of such a sequence: **ACCCGTCCGTGTTAG**. You can think of these nucleotides as the DNA alphabet, and the entire genetic code as a book. Unlike the books we normally read, the DNA text is an *uninterrupted sequence* of letters — a triplet of bases codes for one of the many amino acids (building blocks of protein) needed for an animal to develop, live, and reproduce.

It will be immensely helpful if you picture an organisms DNA as a one-of-a-kind book — unique and special in each individual. Each chromosome can be compared to a chapter segmented into many paragraphs. Each gene occupies a sentence within a paragraph.

Scientists who "read" this genetic "book" find nonsensical gibberish written between the sentences of each paragraph. When they finally identify the actual "sentence" that defines a specific gene, they've hit paydirt. In essence, a gene is a sentence in the cell's book of inheritance. Just as in a real book, DNA can only be read in one direction. Imagine the wonder of the DNA alphabet only consisting of four letters! Think of the many different books that nature writes with this simple quartet!

Each gene pair is singly or as part of a group responsible for a specific trait or traits in the organism it occupies.

Before you decide to have children, inspect the ears of your prospective mate. The trait for brittle earwax is inherited in a simple *Mendelian* fashion by a single *gene pair* at one *locus* (genetic "address") — along with the trait for tufts of hair growing on ears, counterclockwise cowlicks in hair, attached earlobes, and other human physical oddities. The choices at these loci are **brittle wax-normal wax, ear tufts-no ear tufts, attached lobes-unattached lobes** and so on. Many conformational traits, however, result from multiple (poly) genes working together. Such genes are usually not located close to each other — sometimes they're not even on the same chromosome.

It's important to remember that genes do not mesh or blend together. In kitchen vernacular, DNA is not a puree but a fruit salad. Some genes suppress the expression of others, some band together with others to express a trait, some mutate, some even jump location from one chromosome to another — they are indeed a lively bunch! Nevertheless, each remains a separate entity responsible for a specific trait or function. This is the reason that two genes or alleles (alternate forms of a gene) — one inherited from the sire, the other from the dam — <u>separate themselves again in subsequent generations</u>.

You will understand the concept better when we discuss dominant and recessive alleles later on. A good example is the recessive allele for *albinism* in humans and in other mammals. Its *alternate* dominant partner is *no albinism*. After you read about dominant and recessive traits, you will understand how a trait such as albinism surfaces in a population.

Finally, let's not lose sight of the purpose behind all this packing and unpacking, duplicating and repairing, meeting and departing. The cell's activities are cleverly designed to ensure the continuation of the species. Creating new life and supporting the individual organism is Nature's grand scheme to secure DNA's immortality.

A llama's legs are made of protein, not DNA, although DNA determines their length, shape, thickness, and fiber coverage. To understand how it happens, let's dig up a little sports history (stop groaning about my analogies — paring this stuff down to a basic level is no picnic!). If you're a baseball fan, you are probably aware of the most famous trio in the sport. *"Tinker to Evers to Chance"* became as common a phrase for smoothly executed teamwork as *"Mother, Home, and Apple Pie"* did for The American Dream. The cell's well-rehearsed and superbly performed genetic double-play is *"DNA to RNA to protein."* Let's describe and admire its unparalleled precision.

First, DNA is *transcribed* into an RNA (ribonucleic acid) molecule. You may want to picture this as university English (DNA alphabet) copied and transcribed into Southern dialect (RNA alphabet). Actually, three different types of RNA are involved in this process.

During the second step, the appropriately named *messenger* RNA *translates* RNA into protein (like translating English into German). Unlike the DNA and RNA alphabets, which contain four "letters" (the nucleotides), the language of proteins has twenty (amino acids). In effect, the cell's "dictionary" translates a four-letter DNA alphabet into a twenty-letter protein alphabet. The somatic cells have to know the specific protein they're producing and when it's time to stop or re-start making it.

If I've whetted your appetite for more detailed information about the molecular process of genetics, I suggest you read *The Nature of Life* (1989), by John H. Postlethwait and Janet L. Hopson. Their book is clearly written, and large, colorful illustrations make the study of its contents a pleasure.

So, Dear Reader, when contemplating genes, think *suitcase*, *ladder*, or *floss-toy*. Think *alphabet*. Think *Tinker to Evers to Chance*. Think of that little mechanic efficiently repairing most errors.

How many genes do humans or other mammals have in their genome? I've read figures as low as 30,000 and as high as 120,000 for humans. Numbers also vary considerably from species to species.

As I've said, you need a nimble mind to keep up with these *workers*, as research scientists like to refer to themselves.

Chapter Five

SNIPS, SNAILS, AND PUPPY DOG TAILS

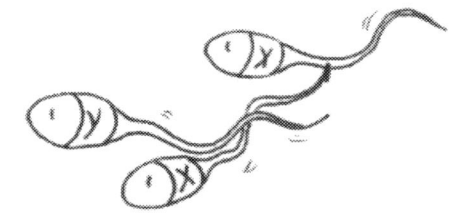

CHAPTER FIVE

Snips, Snails, and Puppy Dog Tails

"Chromosomes: in animal cells, a structure in the nucleus containing a linear thread of deoxyribonucleic acid (DNA), which transmits genetic information ..."
Blood & Studdert, *Baillière's Comprehensive Veterinary Dictionary*, 1988

Previously, we discussed the number of chromosomes specific to each species. It's time to study the concept of *sex chromosomes* versus *autosomes* in more detail.

Normally, all mammals have two sex chromosomes. The remaining chromosomes are *autosomes*. You may see the word *autosomal* precede the terms *dominant inheritance* or *dominant traits*. A writer might say, "This trait is passed on in an autosomal dominant mode." It simply means that a gene can be transmitted to either sex. An example is the allele coding for black coat color in dogs, or the allele producing red (fawn) fleece color in camelids.

A female normally has two identical sex chromosomes, making her **XX**. Males normally carry one **X** and one **Y**. The **Y** chromosome is strictly a "guy thing." I use the word "normally" advisedly in the explanation of **X** and **Y** because there are always exceptions. That seems to be the case with anything pertaining to genetics! In birds, for example, it's the exact opposite — female birds carry the heterozygous sex chromosome combination.

Very little genetic material is carried on the **Y** chromosome. It is also noticeably smaller than its counterpart. Scientists have observed that **Y**-bearing sperm swim faster than their **X**-bearing competitors. No wonder! The little bums travel light. They don't have the weighty responsibility of being genetic beasts of burden! I am kidding, of course. While the **Y** chromosome does take it easy, the other chromosomes in the sperm cell carry the same genetic load as their egg cell counterparts (although you may recall that the sperm cells do not contribute mitochondrial DNA to the zygote).

In the world of DNA, the female cells are truly interested in being good sports and keeping things on an even keel. Several weeks after fertilization, each embryonic cell belonging to a female organism generously "shuts off" all activity on one of its **X** chromosomes. The genes on that particular chromosome thus become inactive, forming what geneticists call a *Barr body*. All daughter cells dutifully follow the mother cell's orders. The process of in-activating one **X** does not select one over the other. Some cells inactivate the sire's **X** chromosome, others the dam's. Scientists therefore refer to female mammals as "genetic mosaics." Yes, that also includes women!

There are exceptions. In the cells of the "...extraembryonic lineage, such as the trophoblasts of the placenta and the visceral endoderm of the yolk sac...the paternal genes are silent" (Marisa S. Bartolemei and Shirley M. Tilghman, "Genomic Imprinting in Mammals," *Annual Review of Genetics*, Vol. 31, pp 493-525, 1997).

For roughly six weeks after conception, human male and female embryos are biochemically identical. Until that time, hypothetically speaking, the cells that slowly develop in the lower body of the fetus can become the reproductive organs of either sex.

In *The Secret of Life* (1993), Joseph Levine and David Suzuki vividly describe what happens at that point as a "chemical messenger ... travels through the body and turns on certain genes in several different tissues." *Testes* form in the male and *ovaries* in the female. The stage is set for further (and future) development of *sex limiting* characteristics (penis, breasts, etcetera), and the eventual birth of the boy or girl, bull or heifer, dog or bitch. The timeframe for this type of development varies, of course, with each species.

Since the male of the species is the one carrying the **Y** chromosome, his genes alone determine the sex of the offspring. Think of the guilt, shame, and embarrassment that has overwhelmed millions of women over the centuries when they gave birth to unwanted daughters instead of wanted sons — and the verbal and physical abuse that many endured because they did not produce a precious heir in male-oriented societies. Queens lost thrones (indeed, some even lost their heads) over their presumed inability to produce heirs, when all along the King's **Y**-bearing sperm cells were having a slow day.

What about the inevitable exceptions mentioned earlier? Sometimes, because of yet undetermined reasons, chromosomes are mixed up or genes become defective during the development of an embryo. Researchers report that among violent criminals there are many men who carry an extra **Y** chromosome, making them **XYY**. Males may be **XXY** as well. There are other irregular combinations such as **XXX** or **XO**.

According to Levine and Suzuki, phenotypically normal women can be **XY**. They carry testes in the space normally reserved for the female repro-ductive organs. These women, through a quirk of nature, cannot reproduce, but otherwise they look like and consider themselves to be females. They do not experience identity crises, nor do they wish to be men. Levine and Suzuki's conclusions conflict with information given by other authors that the presence of the **Y** chromosome always results in a phenotypical male in mammals. The choice of the word "always" is a dangerous one to use in the field of genetics.

Levine and Suzuki describe the predicament of some of the women athletes banned from participating in the Olympics after "failing" DNA tests (named SRY tests for the sex-determining region of the **Y** chromosome) imposed by the International Olympic Committee. They explained that some **XY** women, like Spanish athlete Maria Patino, do not carry the mechanism for "turning on" the bodies' ability to detect and utilize the testosterone naturally produced by them. Therefore, they have absolutely no athletic advantage over **XX** women. The authors quote Peter Goodfellow, a British scientist. While discussing the subject of using SRY tests on athletes, Goodfellow declared: "I think it's an example of woolly thinking. They haven't defined what it is they're trying to ascertain by using SRY for sex tests. A good example of the problems would be **XY** women who have mutations in SRY. They would score as positive [male] on this test, but they are women. They have no advantages over any other woman when it comes to athletics." Cases such as Maria's are sad examples of how people can abuse their power by mis-interpreting genetic facts and applying complex test results that they do not truly comprehend.

Abnormalities involving the sex chromosomes are much more common in humans as well as animals than non-scientists suspect. The overwhelming majority results in sterility.

Three decades ago, I came across a canine hermaphrodite (an individual having both male and female reproductive organs) in the kennels of a German breeder of Afghan hounds. I have not seen another, although I imagine that many breeders would just quietly cull such an animal and not acknowledge its existence publicly. Denise spoke to an owner whose Greyhound bitch was documented by veterinarians as a true hermaphrodite, and a paper will be written and published about her case.

The birth of twins can result in great joy to the breeders of livestock. Sheep breeders select for the trait of multiple births, with breeds such as Finn sheep producing "litters" of five or six lambs. Cattle farmers are not so happy when twins born to their cows are male and female — invariably the little heifer (female) will be sterile. In utero, the placental blood vessels fuse, and the testosterone secreted by the male does not permit normal development of the female organs. Such calves are called *freemartins*.

In insects, we have sexual oddities called *gynandromorphs*. These animals have half-male and half-female bodies. The two halves are phenotypically distinct and make the wasp or butterfly look like someone had cut two different insects apart lengthwise and glued the two mismatched halves together.

Most breeders of livestock and even pets naturally cheer for the **X** chromosome to beat out the **Y**. The commercial livestock industry has little need for large numbers of intact (un-neutered or un-gelded) males. The sperm from one bull is sufficient to artificially inseminate thousands of cows.

Breeders of species not destined for slaughter should welcome the birth of males as joyfully as the appearance of females. Healthy genetic diversity cannot exist in a species or a breed using a very limited number of studs. Those boys are needed! I reminded myself of this when I witnessed the birth of our first male cria, Stormwind's Gregor Mendel. His half-brother, Stormwind's Sir Francis Crick, arrived a few months later. There is nothing like putting your money where your mouth (or pen) is!

When genes are exclusively carried on one of the sex chromosomes, we describe them as *sex- linked*. An example is the tortoiseshell pattern in cats. A *sex-linked* color gene is very rare in mammals. In cats, a mutated gene for yellow coat color is carried on the **X** chromosome. Females who are homozygous for this color are yellow. When bred to a black male, the male off-spring are yellow and the females are tortoiseshell. In *General Genetics*, the authors list the following choices:

<table>
<tr><td>Females</td><td>Males</td></tr>
<tr><td>**yy** = yellow</td><td>**y(-)** = yellow</td></tr>
<tr><td>**Yy** = tortoiseshell</td><td>**Y(-)** = black</td></tr>
<tr><td>**YY** = black</td><td></td></tr>
</table>

The lower case **y** represents the gene on the **X** chromosome that codes for yellow coat color; the upper case **Y** denotes black. The authors are careful to point out that the **Y** does not stand for the **Y** chromosome in this case.

Tortoiseshell males will rarely make an appearance in the kitten nest. They are usually sterile. We can speculate that such males may carry an extra chromosome — they're **XXY**. Interested cat breeders can study a very detailed account of this phenomenon in *The Nature of Life*.

The most famous example of a sex-linked trait is the white eyes of the Drosophila fly. They first appeared as a mutation on the **X**-chromosome of a male. The original (wild type) Drosophila had only red eyes.

In two-legged mammals, the gene for channel-surfing is exclusively carried on the **Y** chromosome.

This concept is not to be confused with the *sex-limited* gene action involving traits that can be <u>carried</u> by both sexes but are only <u>expressed</u> by one. A stud may carry the gene resulting in *agalactia* (poor milk production). As a male, he cannot express the trait but can certainly pass it on to his daughters. In alpacas, it is suspected that fused ears are inherited as a sex-limited trait, with only females expressing the defect.

Dr. Willis presents a fascinating example of a sex-limited genetic mechanism. Ayreshire cattle come in two colors — mahogany (**M**) and red (**m**). Obviously, cattle homozygous for mahogany (**MM**) will only express that color. All animals homozygous for red (**mm**) can only be red. Simple! Now it gets interesting. The heterozygous combination (**Mm**) produces mahogany coats in males and red coats in females. These red females are genetically **Mm** and reproduce as such.

In a previous chapter, you learned about mitochondrial DNA. Some breeders erroneously assume that mitochondrial DNA is carried on the **X** chromosome. It is not! Remember that it is located in the cytoplasm of the cell. Males as well as females inherit mitochondrial DNA only from their dam. Both males and females have mitochondrial DNA in their somatic cells. The female gamete (egg) also has mitochondrial DNA, but the male gamete (sperm) does not. Since the sperm cell does not contribute cytoplasm to the zygote, the sire's mitochondrial DNA is not passed on to the offspring. There is much speculation about the influence of mitochondria on performance traits, such as speed in Greyhounds.

You might come across the term *sex-conditioned character*. Human baldness serves as a good example to explain this concept. In females, the condition is often expressed to a minor degree (very thin, sparsely growing hair). In males, we find full expression (and who the heck knows why they're so sensitive about that subject).

Sometimes people think of certain personality traits as sex-limited. For example, the perception among the general public is that bitches make better pets than males. In my own breed (Whippets), I can't agree with that premise. The consensus among long-term breeders we've spoken to who own both sexes is that "the boys" are the more affectionate and biddable of the two. Denise agrees with me, at least in the case of un-spayed or "intact" bitches — without estrus cycles to constantly disrupt hormone levels, she usually finds the males more laid back than "the girls." Not always, but often.

From my own observations, alpaca males seem to be more people-friendly than the females. Although advertised as "huggable investments" by many breeders, pregnant females appreciate being hugged about as much as the average fourteen-year-old boy likes being kissed in public by his mother. Our gelding actually enjoys being smooched, but only if none of the other alpacas are loitering around the barn to watch.

In the next chapter, we can totally forget about kissing, hugging, or any other mushy stuff. It's fighting and feuding and — oh, you'll see!

Chapter Six

GENOMIC IMPRINTING

CHAPTER SIX

Genomic Imprinting

"A handful of autosomal genes in the mammalian genome are inherited in a silent state from one of the two parents and in a fully active form from the other."
Dr. Bartolomei and Dr. Tilghman, *Genomic Imprinting in Mammals*, 1997

When my brother heard I was writing a book about genetics, he immediately asked, "Does it have sex and violence in it?" Then he added, "That's the only thing that sells, you know!"

I told him how ignorant he was. "Of course the subject of genetics is chock full of sex and violence."

"The sex part I understand," he conceded, "but violence?"

I urged him to pay attention. "Listen to this! Tug-o-war ... weapons are imprinted genes ... epigenetic arms race ... conflict ..." (*Genomic Imprinting in Mammals*, 1997).

I was feeling righteous. "What do you have to say now?"

Impressed, my brother was momentarily as silent as some genes that mammalian species either inherit from their sire or their dam.

Scientists did not discover *genomic imprinting* until the early 1980's. Until then, the sex of the parent was not believed to play a role in determining the expression of a genetic trait. That concept actually still applies to most of the genes carried by mammals. If the gene for red fleece color (which is dominant in camelids) is inherited by a llama, it does not matter whether the **A** allele came from the sire or the dam — the llama will be red (brown).

This genetic action also holds true for *polygenes* and *pleiotropic genes* (more about these later). We can say that both parents have equal

41

opportunities to see their traits inherited by and expressed in their offspring — but not always!

"A handful of autosomal genes in the mammalian genome are inherited in a silent state from one of the two parents, and in a fully active form from the other, thereby rendering the organism functionally hemizygous for imprinted genes" (*Genomic Imprinting in Mammals*).

Many details in the *Annual Review of Genetics* (Dr. Marisa S. Bartolomei and Dr. Shirley M. Tilghman) that follow this introductory abstract are much too complex for the layperson to comprehend. I will try to present an understandable and simplified framework for this complicated concept.

Let's return to two key words here. *Hemizygous* means that the gene is present in a single dose, such as happens in a haploid cell. Notice that in the previously mentioned article the word hemizygous is preceded by the word *functionally*. This goes right to the heart of the matter. Two alleles are present, but only one of them is in working order — the other allele is silent.

When dealing with imprinted genes, scientists speak of *maternally* (from the female parent) or *paternally* (from the male parent) *expressed* genes. This can be confusing to breeders. We think of a gene as expressed in the offspring, such as the **A** allele I mentioned earlier. For example, if a cria is **Aa**, it carries a recessive allele for black, but it only expresses the dominant **A**.

When you study the concept of genomic imprinting, you need to put the use of the word "express" out of your mind in connection with "offspring."

In that context, you must think of genes as expressed from the maternal or paternal chromosome. A *paternally expressed gene* refers to, for example, the mouse gene **Igf2**, which is only active in the offspring when it is inherited from the sire. The **Igf2** allele inherited from the dam is always silent. To quote Dr. Tilghman (from personal conversation): "The sex that matters is the sex of the parent, not the sex of the offspring." Both male and female offspring carry and show the effects of paternally as well as maternally expressed genes.

Do not confuse genomic imprinting with recessively inherited genes. Recessive alleles are also "silent" until they appear in homozygous form. Homozygosity is not a prerequisite for the function of imprinted genes. (*Homozygosity*: the state of having identical alleles in regard to a given character or characters — D.C. Blood and Virginia P. Studdert, *Bailliére's Comprehensive Veterinary Dictionary*, Bailléire Tindall, 1988.)

Genomic imprinting might be easily confused with the idea of sex-linked traits residing on the **X** chromosome. In a broader sense, the inactivation of one **X** chromosome in all somatic cells can be considered an imprinting mechanism, but let's not forget that it happens randomly. The one exception, as I mentioned in a previous chapter, involves cells of the extraembryonic lineage, which exclusively shut off action in the paternal **X** chromosome. A slightly <u>modified</u> version was found in marsupials, that inactivate only the paternal **X** chromosome in <u>all</u> somatic cells (remember that this normally happens randomly in mammals). It is important to understand that most imprinted genes discovered so far do <u>not</u> reside on the **X** chromosome.

In *Genomic Imprinting in Mammals*, the authors present a table of the nineteen imprinted mouse genes that were found up to that date (1997). Only one (**Xist**), a paternally imprinted gene, is located on the **X** chromosome, ten (called a cluster) on chromosome **7**, two on chromosome **17**, and two on chromosome 2. **Xist** is the only gene activated from the otherwise silent paternal **X** chromosome in the extraembryonic lineage cells — making it the exception to the exception. Oh boy!

Thirty-four imprinted genes were identified by July 2000, most of them in mice and humans, one in the opossum. The latter is classified as a marsupial, one of the lower mammals that lack a placenta. The prematurely born baby is carried to full development in its mother's pouch.

In "The Sins of the Fathers and Mothers: Genomic Imprinting in Mammalian Development," Dr. Tilghman says that "two themes emerge" after searching for a central role that imprinted genes might play in the development of mammals (*Cell*, Vol. 96, 185-193, January 22, 1999). Many imprinted genes influence fetal growth. Others play a role in brain development.

Exactly which traits are under such genetic control?

Brace yourselves. This is where it gets violent! The battle of the sexes, as we are about to discover, unfortunately starts in the womb. The theory is that in promiscuous mammals, the father, concerned more about his babies than the health of their mom, tries to genetically "optimize the reproductive fitness of his offspring by promoting their growth even at the ex-pense of the mother's future litters" (Bartholomei and Tilghman). We find **Igf2** in his arsenal.

Mom, on the other hand, realizing that her body must conserve resources and save energy for future babies sired by other males, tries to inhibit growth and arms herself with **Igf2r** that directs the degradation of **Igf2**. Extensive research proved the effect of the paternally expressed mouse gene **Igf2** on fetal growth in that species. When scientists purposely mutated this gene (*targeted mutation*) to a "loss of function" form — meaning the gene was essentially deleted — the resulting mice were 40 percent smaller than normal.

So, the weapons (imprinted genes) are drawn and the shootout at the DNA corral proceeds, embroiling the embryo in what one scientist describes as a "parental tug-o-war." It is interesting to note that scientists trying to grow viable mammalian organisms by using the combined nuclei from two male or two female cells did not succeed.

What chance do the sexes have to live in complete harmony with each other when such bloody conflicts are already taking place between the paternal and maternal genes in the cell? Who knows how many genes will eventually be discovered in this genetic "War of the Roses"? Dr. Tilghman thinks possibly 75 to 100.

Genomic imprinting might also account for another well known but little understood phenomenon. During the history of mammalian evolution, it occasionally happened that a species split into two populations that lived geographically apart for thousands of years. When members of such populations are reunited and bred, they are often unable to produce viable or fertile offspring. It is now believed that in such cases imprinted genes have diverged to the point of rendering members of the previously separated species "reproductively incompatible, leading to speciation" (*Genomic Imprinting in Mammals*).

Apparently, imprinted genes vastly complicate the matter of cloning, and are implicated in the failure of many cloned embryos to develop to full term. Dolly, cloned from the mammary cell of a mature ewe, was the only survivor of 277 reconstructed embryos. Although scientists obviously surmounted the problem of cloning an animal from a *differentiated* (somatic) cell (a feat previously thought impossible), they are not yet able to <u>routinely</u> prevent *large-fetus syndrome* (among other problems).

Embryos constructed by nuclear transfer are often so large that a normal birth is impossible. When the American cattle industry began to clone, the breeders did not count on the high veterinary expenses — caesarians are not cheap — caused by large-fetus syndrome. The disruption of normal development of imprinting routines during cloning is implicated in this problem.

The work on cloning continues. The March 2001 issue of *Lancaster Farming* featured the two calves cloned from cells "donated" by Zita, the Holstein cow belonging to the Wiles family of Williamsport, Maryland.

The full story of genomic imprinting is yet undiscovered.

Breeders should be careful not to jump to hasty conclusions. There is presently no evidence that imprinted genes code for important conformation or performance traits (such as speed), as one article I read suggested. Also remember that probably <u>all</u> members of a species or breed will carry imprinted genes specific to that population. If superior racing performance in horses was based on a maternally expressed gene, for example, <u>none</u> of the excellent stallions could pass his racing ability to his immediate offspring.

Meanwhile, just because some male and female genes are locked in mortal combat does not mean that we and our spouses, partners, friends, co-workers, and siblings cannot get along and cooperate. Let's rise above our violent genetic heritage and —

"Oh, for Pete's sake," my brother interrupted with a groan. "Get off your soap box! What kind of weird book is this? You sound like a television preacher, not an author."

There he goes — the recipient of an imprinted personality gene — paternally expressed. On reflection, I must admit that I probably inherited it myself. Remember, it is the sex of the parent that counts, <u>not the sex of the child</u>!

New discoveries are found in the field of genetics each day. The purpose of this chapter was to introduce the reader to the <u>concept</u> of genomic imprinting. For updated information on newly discovered imprinted genes, interested readers should follow updated research.

Chapter Seven

DOMINANT GENES

CHAPTER SEVEN

Dominant Genes

"Dominant gene: one that produces an effect (the phenotype) in the organism regardless of the state of the corresponding allele. Examples of traits determined by dominant genes are short hair in cats and black coat colour in dogs."
Blood & Studdert, *Baillière's Comprehensive Veterinary Dictionary*, 1988

A "dominating woman" ... "he is the dominant one in the family" ... "her mother-in-law wants to dominate her life" (mine didn't, she was adorable) ... If a room is primarily decorated in red, we describe the red color as "dominating the space." We use the word *dominant* often.

Webster's Dictionary defines **dominant** as 1. *dominating; ruling; prevailing; exercising authority or influence.* There are dominant whites in horses, yet they are relatively rare in that species. Black coat color is dominant in Whippets and other dog breeds, yet you will not find many black Whippets in the show ring. They are more popular on the race track or lure coursing fields, or even as pets, but twenty years ago you would have been hard-pressed to find one at all. Are you confused?

Don't let this puppy's sweet expression fool you. Young Whippets are very active and need lots of exercise and attention (photo by Barbara Ewing)

The concept of *genetic* dominance is easy to understand once you realize that it does not necessarily mean "more of." It should shortly become clear that a breeder can make a *dominant* trait disappear with almost the same ease as a skilled magician puts the rabbit back into his top hat. In our case, however, the "rabbit" is gone forever.

The *phenotype* of an animal is plain for all to see. Conformation, coat- or fleece quality, and color are easily evaluated by visual observation and by actually touching the animal (in the case of coat- or fleece texture). What we do not see are the hidden *recessive* traits. They surface only when two recessive genes (alleles) are paired together, one from the sire (father) and one from the dam (mother). We refer to the entire package, the genetic make-up of an animal, as its *genotype*.

Let's use Whippet color to explain. Let's say that several black as well as several red specimens of these speedy little sighthounds reside in our (fictitious) kennel. It is customary to use a capital (upper case) letter to indicate dominant traits, and a small (lower case) letter for recessives. If an animal carries both (heterozygous), the dominant gene (allele) is always listed first.

A BREEDER'S GUIDE TO GENETICS
Relax, It's Not Rocket Science

In our example, **A** represents the black allele and **a** represents the red alternate (its diluted form — fawn — is a very popular color among dog show enthusiasts). We have:

        Felix **AA** — phenotype black
        Gladys **AA** — phenotype black
        Ayla **Aa** — phenotype black
        Cleo **aa** — phenotype red

Ayla demonstrates how, whenever a dominant and recessive allele are paired, the dominant one "overpowers" and prevails over the recessive one. The phenotype of that animal is thus fixed. Let's see what happens when Felix, our energetic young stud, is mated to each of the three bitches. We will use what we call a *Punnett Square* (named after Britain's Reginald Crundall Punnett) to help us figure out the genetic *possibilities*.

You've learned that, in many species, females are **XX** and males **XY**. The following diagram will help you to understand how to interpret the Punnett Square (also referred to in textbooks as a *grid* or *checkerboard*). I will use the sex chromosomes to explain.

		sire	
		X	Y
dam	X	XX	XY
	X	XX	XY

Two of the offspring are females. Two are males.

The dam can only give an **X** to her offspring. The sire, however, can give **X** or **Y**. Remember that <u>each</u> parent only passes on <u>one</u> chromosome in a homologous pair. Each parent can only pass on the genes (alleles) <u>located on that particular chromosome</u>. The "partners" of these genes (alleles), residing on the corresponding chromosome, are not inherited by the offspring.

While this example teaches you how to read a Punnett Square, it does not explain the concept of a dominant gene. For that purpose, let's move on to our breeding program.

Felix bred to Gladys:

	A	A
A	AA	AA
A	AA	AA

All the puppies are black.

Felix bred to Ayla:

	A	A
A	AA	AA
a	Aa	Aa

All the puppies are black. Two puppies carry a red recessive.

Felix bred to Cleo:

	A	A
a	Aa	Aa
a	Aa	Aa

All the puppies are black. All four carry the recessive for red.

The breeding program does not stop here. We find Rocky, a beautiful red (**aa**) stud, at another kennel.

A Felix x Cleo daughter is bred to a very happy and willing Rocky.

	a	a
A	Aa	Aa
a	aa	aa

Two puppies are black — they also carry the recessive for red. Two are homozygous for red.

The black allele, although dominant, never made an appearance in the genome of the two **aa** puppies. Our two red puppies can only produce black offspring in future litters if they are bred to a black mate.

You will learn that the inheritance of color is determined by more than one locus. In that respect, my presentation here is somewhat simplistic. The objective of this chapter, however, was to clarify *Mendelian inheritance*.

Remember that in other species the dominant **A** might very well represent *red* and the recessive **a** *black* coat or fiber.

You must keep one very important point in mind: Whether an **Aa** animal passes on **A** or **a** to his/her offspring (geneticists use the term *progeny*) is left to chance and the law of probabilities. If you paid attention when your 4th grade teacher explained this during math class, you will know what I'm talking about. If you were busy throwing spitballs, you should listen up now!

Chapter Eight

GENETIC PROBABILITIES

CHAPTER EIGHT

Genetic Probabilities

"Probability: the basis of statistics. The relative frequency of occurrence of a specific event as the outcome of an experiment when the experiment is conducted randomly on many occasions ..."
Blood & Studdert, *Baillière's Comprehensive Veterinary Dictionary*, 1988

Let's pretend you have a brown bag in front of you, filled with ten marbles. Six of the marbles are black, four are red. Your *choice* of color is 50-50, either black or red. The *probability* of choosing black is 6 out of 10 or 60 percent; the *probability* of choosing red is 4 out of 10 or 40 percent. Each pick is a separate entity and does not affect any other picks as long as the original contents of the bag are restored each time. You may wonder what this has to do with breeding animals.

Let's revisit our kennel. This time we plan to breed Ayla (**Aa**) to Bogie (**Aa**). As you can see, both express the black phenotype.

		sire	
		A	a
dam	A	AA	Aa
	a	Aa	aa

Your *choice* of color is either black or red.

Three puppies are black — two of those three puppies carry the recessive for red.

Now comes the clincher! The fourth puppy is red and — *abracadabra* — the <u>dominant</u> black allele, <u>expressed by both parents</u>, has <u>disappeared</u> in that offspring. Gone! Never to be seen again!

Expressed as *probabilities* of phenotype and genotype, the litter would be neatly divided into 25 percent black (**AA**), 50 percent black (**Aa**), and 25 percent red (**aa**). Let's say you want black puppies out of this breeding. You are thrilled to figure out on paper that 75 percent will be black, and only 25 percent will be red. What is wrong with this picture?

Actually nothing, as long as you realize that we are speaking in *probabilities* only. In fact, this breeding might produce 10 red (**aa**) puppies. The odds are against that happening, but genetically it is possible.

Remember when you played Parcheesi when you were a kid, and you needed to roll a "six" to move your piece on the board? The *probability* of you doing that was one out of six. Do you recall the times you had fits because after twenty rolls the six had still not come up? That's because mathematical ratios (in the case of our litter 25-50-25) as predicted by the *Theory of Mathematical Probabilities* are most closely reflected in reality when large numbers of events are examined.

Breeding is a little bit like rolling the die, unless both sire and dam are *homozygous* (*homo*=same) for the same trait. If both animals are **AA**, then it follows, as we saw with the Felix x Gladys breeding, that all the progeny must be black. If both animals are homozygous for red (**aa**), then all the puppies will be red. It is only when both parents are *heterozygous* (*hetero*=different), or one parent is heterozygous and the other is homozygously recessive, that the phenotype can be either black or red.

A BREEDER'S GUIDE TO GENETICS
Relax, It's Not Rocket Science

		sire: heterozygous	
		A	a
dam: homozygously recessive	a	**Aa**	**aa**
	a	**Aa**	**aa**

Breeders often labor under the misconception that a heterozygous animal always passes on the dominant allele (it's that unfortunate choice of the word "dominant"). As I have shown, this is not true. Only *if* it is passed on does it overpower its recessive partner.

Picture the two alleles as wrestlers. The dominant one is a 250-pound heavyweight. The recessive one only weighs 103 pounds. When they wrestle, the little guy does not stand a chance. We can't even see him as the heavyweight has him pinned to the mat. As they leave the gym, however, they're an even match. Some days the heavyweight beats the lightweight to the door, and on other days the little guy sprints past his opponent and leaves him behind.

Considering this information, let us reexamine my statement that black is the dominant allele in Whippets, but that black Whippets were extremely rare only a few years ago. Why? For several decades, the overwhelming majority of breeders did not care for this coat color and selected against it in their breeding programs. <u>Removing a dominant gene</u> from a breeding program is relatively simple — <u>do not breed animals that exhibit the dominant trait!</u> Remember, however, that once a dominant allele is "gone," it cannot surface again. Whippet breeders achieved rapid results in removing black coats from their lines.

Now that the color is more acceptable again, breeders are only able to produce black puppies because a few independent thinkers prevented this dominant trait from disappearing all together.

Not all breeders have the choice, and they probably would not want to. All Chesapeake Bay Retrievers, for example, are homozygously recessive at the **B** locus, making all Chessies **bb** (liver). The dominant **B** (black) allele simply does not exist in this breed any more, and cannot be "retrieved" (no pun intended).

I should explain here that the **B** locus does not <u>initially</u> determine whether a dog is black or not. Its only function is to code for black or liver once a dog is **A-**. An **aa B-** dog cannot be black. In camelids, where black is believed to be <u>recessive</u> to red, the **A- B-** alpaca, for example, cannot be black because the **A** allele codes for red at a separate locus.

Incidently, blue Whippets — like blue dogs of other breeds — are genetically black. When breeders ignorant of color genetics tell you that two blacks did not produce any black offspring, you will often discover that one or several puppies are blue (**A- dd**) — diluted blacks. You will learn in Chapter Thirty how specific genetic combinations such as **dd** dilute ("wash out") the color of the basic pigment.

To complicate the issue I must tell you that in certain dog breeds (such as the Chow Chow), two red parents may very well produce black offspring, despite the fact that black is dominant. You probably won't understand that completely until you've read the chapters on color inheritance. The genetic formula for such an occurrence is **AA ee** x **aa EE** = **Aa Ee**. This means that two genetically different "reds" exist in that breed. The **A- ee** Chow Chow is actually a "disguised" black.

Don't fret about this color inheritance preview now. Simply tuck the information in the back of your mind and refer to it again after you have read Chapters Thirty and Thirty-one.

While most dominant genes code for *functional* traits, there are a few nasties that code for *loss of function* in an extremely sneaky manner. Read on.

Chapter Nine

LETHAL DOMINANTS

CHAPTER NINE

Lethal Dominants

"<u>Lethal</u>: deadly;fatal. <u>Lethal trait</u>: an inherited characteristic that ensures the early death of the inheritor. <u>Lethal white</u>: the name given to the progeny of white horses in which the gene for the white character is dominant. Their progeny have a very high mortality rate. The same applies to the offspring of a mating of two Overo horses; the defect in these is atresia of the intestine."
Blood & Studdert, *Baillière's Comprehensive Veterinary Dictionary*, 1988

Besides the "<u>once you lose it, it's gone forever</u>" rule of the dominant allele, white horses are rare for another reason. Only heterozygous white (**Ww**) horses are born alive. If a foal zygote has the misfortune of receiving a dominant white allele from both parents, making it homozygous (**WW**) for white, it will die *in utero* (in the womb). In the complex world of color genetics, **W** denotes the dominant white allele (other genes also produce completely white animals, but we will discuss them later). The single lower-case **w** tells us that the animal carries a recessive allele permitting color to express itself. The **ww** informs us that the animal has a colored coat or fleece. <u>The color is determined by other genes</u>. The choices at the **W** locus consist only of **dominant white - not dominant white**. White markings are produced by separate genes and do not come into play here. (The chapter on color will explain the "other" whites to you. They are, genetically speaking, colored.)

If you understood the chapter on dominance, the concept of lethal genes will be easy to comprehend. **Ww** bred to **Ww** can produce three possible genotypes:

		sire	
		W	w
dam	W	**WW**	**Ww**
	w	**Ww**	**ww**

Let's pretend that the four combinations (probabilities) in the Punnett Square are the actual results of four breedings. One foal will be homozygous for the absence of dominant white (**ww**), two will be heterozygous for white (**Ww**), the fourth one (**WW**) will have died sometime before birth. The embryo could have been reabsorbed before the breeder realized the mare was pregnant. As a result, **WW** (homozygous) horses are extremely rare. Dominant white horses are usually **Ww** and therefore do not always breed true. Please remember that **WW** is not necessarily lethal in other species.

The same sad fate befalls the foal embryos that are homozygous for the *roaning* pattern. *Roan* describes an evenly distributed mixture of colored and white hairs. Breeders of other species know of or suspect the lethal effects of certain gene combinations as well. Another example is the homozygous combination (**MM**) for the Manx cat.

William D. Stansfield, Ph.D., tells us in *Theory and Problems of Genetics* (1991) that the typical Mexican Hairless (atrichia) dogs are **Hh**, and the coated ones are **hh**. The **HH** combination is lethal. Additional examples of lethals are the creeper gene in chickens and the gene for yellow coat color in mice. There are others. Make it your business as a breeder to identify the lethal genes in your particular species or breed. It will save you much heartache and money.

Huntington's chorea in man is inherited as an autosomal dominant gene. It does not make its presence known until the person affected with it is an adult and may have already passed the defect to his or her children. Lethal dominant genes, by their very nature, need to employ tricks and ruses to "survive" from one generation to the next.

I have already hammered home the fact that every aspect of genetics has exceptions. Anal-retentive types lead rough lives as breeders (in more ways than one, I might add). Remember our two wrestlers? In the next chapter you will learn how the little guy manages to escape the pin and is clearly seen side by side in partnership with his powerful opponent.

Chapter Ten

INCOMPLETE DOMINANCE

CHAPTER TEN

Incomplete Dominance

"A situation in which the heterozygote state is different from either homozygote."
Malcolm B. Willis, *Genetics of the Dog*, 1989

This is an easy concept to understand because each genotype is expressed as a different phenotype. In *The Horse* (1990), L.D. Van Vleck, et al, describes such a situation in the chapter on Principles of Mendelian Inheritance (he calls it *partial dominance*, a good example of how nomenclature can vary slightly between authors). This chapter illustrates how incompletely dominant genes work.

Horses, like many other mammals, carry the dilution loci **C** and **D**. The C locus affects chestnut colored horses in the following manner: horses homozygous for **CC** express the full rich *chestnut* color; those with heterozygous alleles (**Cc**) have the original chestnut color diluted to a color that horse fanciers call *palomino*; finally, those homozygous for the full dilution treatment will have a cream coat and often will have blue eyes (*cremello*). The coat might look white, but genetically is not even remotely connected to dominant white (**Ww**). On a bay horse, this genetic action produces *bay* (**CC**), *buckskin* (**Cc**), and *perlino* (**cc**). Hypothetically, if the actions at the C locus were under <u>complete</u> dominant genetic control, the phenotype of the bay **Cc** horse would remain bay. Since this isn't the case, and each genotype has its own phenotype, we refer to such genetic action as <u>incomplete dominance</u>.

Dominant white horses, by the way, always have dark eyes and pink skin, notably lacking pigment around the eyes, nostrils, and mouth. If you see a *cremello* horse (phenotypically white with blue eyes), you can assume that it is **ww** at the **W** locus.

It is possible for an animal to be simultaneously dominant white and express the effects of the C locus. Even then (remember the lethal white gene), the genotype of such a horse would still carry one allele coding for color (**Ww**). The phenotype would be white.

Knowledge of such genetic action takes some guesswork out of a breeding program. People who know little or nothing about the inheritance of color are invariably most impressed when this tidbit of information is explained to them.

While we are discussing dominance, I should mention the concept of *co-dominance*. In such cases, two dominant alleles are both equally expressed. An example is the AB blood type in humans. Comb shape in chickens also follows the co-dominant pattern. When a chicken with a pea-shaped (**P**) comb is bred to one with a rose (**R**) comb, the combination (**PR**) produces a walnut comb.

With this chapter, we have killed two birds with one stone. Doesn't that sound terrible? Let's say we've slain two dragons with one sword. You have learned the meaning of *incomplete dominance* as well as *co-dominance* and a small but very important part of color genetics. Now you are ready to tackle the next concept involving dominance with hidden surprises.

Chapter Eleven

INCOMPLETE PENETRANCE

CHAPTER ELEVEN

Incomplete Penetrance

"Penetrance: the frequency with which a heritable trait is manifested by individuals carrying the principal gene or genes conditioning it."
Blood & Studdert, *Baillière's Comprehensive Veterinary Dictionary*, 1988

    This phenomenon is a little trickier to understand and complicates the dickens out of a breeding program. It serves, however, as a viable explanation for certain genetic patterns that simply do not make sense to a breeder.

    You learned that when an animal carries either one or two dominant alleles at a specific locus, the trait the dominant allele codes for shows in the animal's phenotype. There are interesting exceptions to such a clear Mendelian inheritance pattern. We'll use "fantasy" colors to explain: both **ZZ** and **Zz** are phenotypically purple, and **zz** is pink. The very term *incomplete penetrance* tells us that some action is happening, but is not being completed. Under such genetic action, the genotype **Zz** would occasionally exhibit the identical phenotype as **zz**. In other words, the genetically purple animal looks pink. The breeder is then confused when this individual passes on the trait for purple in a normal dominant fashion.

Suri alpacas are more rare than their fluffy Huacaya counterparts. Members of both varieties are shorn once a year. Their incredibly soft, luxurious fiber is used to produce garments, blankets, rugs, and craft items (photo by FiberGenix Suris)

In *Genetics of the Dog*, Dr. Willis reminds breeders that they "should bear in mind the existence of this phenomenon ..." if they encounter results in their breeding programs that, at first glance, do not make sense.

Several alpaca breeders have reported the births of Suri crias out of Huacaya parents. The terms *Suri* and *Huacaya* refer to different fleece varieties. I did not see these crias (babies) and therefore cannot vouch for the authenticity of these claims. The allele responsible for the Suri type is normally considered the dominant one. It follows that a combination of a Huacaya (**ss**) bred to another Huacaya (**ss**) cannot produce a Suri.

		sire	
		S	S
dam	S	SS	SS
	S	SS	SS

If the concept of incomplete penetrance applies, one parent would be heterozygous for the Suri trait (**Ss**) without expressing this in its phenotype.

		sire	
		S	S
dam	S showing as s	**Ss**	**Ss**
	S	SS	SS

The Punnett Square shows clearly that the probability of such a mating producing Suri progeny is 50 percent. This explanation is <u>purely speculative</u> at present and should not be viewed as a genetically proven fact in alpacas.

The North American birth rate of Suri offspring from two Huacaya parents, as documented by the Alpaca Registry Incorporated (ARI), is very tiny. Although the genetic model of incomplete penetrance may not apply, the surprising appearance of such Suri babies is surely the result of a specific gene action. I will discuss the subject in more detail in the next chapter.

Let me share another example — this one is well-documented and not just conjecture. The concept of incomplete penetrance is believed to apply to an autosomal dominant gene producing *polydactyly*. Humans as well as animals that are **PP** and **Pp** have more than the normal number of digits (fingers, toes) on hands or feet. In some families, the **Pp** members do not express the trait, meaning that the gene has less than 100 percent penetrance. I read that some South American alpaca breeders consider extra digits a sign of good luck and do not select against them.

Identifying mode of inheritance is extremely important when trying to purge a breed or species of a serious defect. Mastiff breeders were frustrated when *progressive retinal atrophy* (PRA) did not prove to be under autosomal recessive control as it was established in other breeds. Researchers at Cornell University's Baker Institute found compelling evidence during pedigree research indicating that, in Mastiffs, PRA is inherited through a dominant gene with *incomplete penetrance*.

Joan Hahn, the Mastiff breed columnist for the *AKC Gazette*, reported on a test breeding which hopefully will contribute to the eventual elimination of the defect.

When <u>all</u> individuals in a population carry a dominant gene expressing the trait it codes for, it is known as *complete penetrance*.

The Blood and Studdert definition for *incomplete penetrance* already explains that more than one gene can be involved. Dr. Willis also stresses that in many cases of incomplete penetrance, *polygenic* (poly = more than one) rather than simple Mendelian inheritance is involved. The chapter covering polygenic inheritance will make that clear to you. Before we pursue that thought, let us make a quick side tour to discuss the opposite of dominant traits — those under recessive genetic control.

Chapter Twelve

RECESSIVE GENES

CHAPTER TWELVE

Recessive Genes

"Recessive: 1. tending to recede; in genetics, incapable of expression unless the responsible allele is carried by both members of a set of homologous chromosomes. 2. A recessive allele or trait."
Blood & Studdert, *Baillière's Comprehensive Veterinary Dictionary*, 1988

Recessive traits are a different kettle of fish. They can pop up unexpectedly and, depending on goals and expectations, bring much delight or despair to a breeder's life. Unless DNA tests exist to identify animals with carrier status, knowledge of pedigrees (ancestral lineage) is the most effective guard against unwanted recessives.

Recessive genes do not "skip" a generation. Remember that an animal must be homozygous (for example: **aa**) for a recessively inherited trait to express itself — hence the erroneous belief held by many breeders that genes can "skip." Recessive genes can be carried, hidden and unknown, for many generations. Old-timers will usually remember that "so-and-so had that in her line and, oh my dear, you would think this breeder would have gotten rid of that in her line by now."

It is only fair to say that owners new to breeding livestock or pets need to understand that breeding deals with possibilities, not total predictability. Animals are not pieces of furniture made to exact specifications. If you desire total control and order in your life, breeding animals should be on the bottom of your list of life's potential pleasures. Resiliency, determination, and limitless patience are virtues better suited to the lifestyle of a breeder.

Unfortunately, many defects, including those suffered by humans, are carried as recessives. Modern science offers a marvelous tool in the form of DNA testing to help animals and people detect genetic problems. Discoveries in DNA research are growing by leaps and bounds. We will discuss them later in the book.

All recessive genes do not code for defects. Occasionally, a trait previously believed to be a recessive one is not, and the label actually applies to its counterpart.

In *Animal Breeding and Production of America Camelids* (1984), Rigoberto Calle Escobar tells us: "When an alpaca Huacaya (sic) **(H+)** is crossed with a Suri alpaca **(H)** the outcome is a Huacaya offspring, suggesting that it simply means a Huacaya dominance over the Suri (recessive)."

There is plenty of evidence to the contrary! Such a cross does not always produce a Huacaya. Reviewing our Punnett Square, we realize that it could just as easily mean that the Suri is not "pure," but heterozygous for the Suri trait. This is not surprising. According to Escobar, in many South American herds the two varieties are allowed to mingle freely and breed with each other.

Since breeding two phenotypically correct Suris often does produce Huacaya offspring (although the reverse situation is extremely rare), it should not shock us that other South American camelid experts and North American breeders disagree with Escobar's statement. Escobar himself leaves room for doubt by concluding that "the stated hypothesis is only a tentative possibility..." and reserves the right to conduct further research. True, there are many more Huacayas than Suris grazing on pastures all over the world. We already discussed, however, the erroneous belief that "more of" always translates into "dominant." It does not!

The comparative rarity of the Suris can be attributed to their higher mortality rates in the harsh climate of the Andean highlands. South American breeders may have unintentionally selected against the survival trait. The previously mentioned rare occurrence of two Huacayas producing a Suri cria tells us that Suri fiber inheritance might not always follow a straightforward autosomal dominant mode. The opposite (two Suris producing a Huacaya) happens quite often.

When two Suri parents both carry the recessive for Huacaya fleece, the probability of producing a Huacaya (**ss**) *cria* (baby) should be 25 percent. The alpaca industry is relatively young — with imports of unknown parentage and no DNA test available, identifying carriers (**Ss**) is usually not possible.

A BREEDER'S GUIDE TO GENETICS
Relax, It's Not Rocket Science

Why did I write "should be" instead of "is" when discussing genetic probability in the previous paragraph? Scientists can examine mating data and, by using mathematical formulas, arrive at conclusions regarding the expected frequency of an allele in a specific population. If actual mating results differ considerably from formulated expectations, we know that genetic action for that particular trait is not as simple as a single gene coding for it.

Dominant and recessive genes, for example, can often impact each other in fascinating ways.

Phillip D. Sponenberg, D.V.M. Ph.D., contributed a very informative chapter on the inheritance of Suri fleece to Eric Hoffman's *The Complete Alpaca Book*. Sponenberg, basing his hypothesis on data provided by the North American Alpaca Registry, states that "the single-gene theory may not be adequate to explain all cases." Although the allele coding for Suri fiber is generally believed to be dominant, the ARI data showed that "in a highly significant departure from what is expected" too many Huacayas are produced from Suri-Huacaya crosses. (This phenomenon may have contributed to Escobar's belief that the Huacaya fleece is produced by the dominant allele.)

What could be the reason for such puzzling results? Sponenberg offers the possible existence of a dominant suppressor gene creating phenotypical Huacayas out of "genetic Suris." Such a gene suppresses the Suri phenotype. Using the Punnett Square, we can work out the probabilities for such a mating quite easily.

We breed a Suri to a Huacaya. The latter carries the suppressor gene. Let's agree on designated symbols to describe the genotype:

S = Suri allele (dominant)
s = Huacaya allele (recessive)
P = Suppressor allele (dominant)
p = Non-suppressor allele (recessive)

Our theoretical breeding is: **SS pp** x **ss Pp**.

If we <u>disregard</u> the suppressor gene, breeding a homozygous Suri (**SS**) to a Huacaya (**ss**) should produce 100 percent phenotypical Suri crias.

The presence of a suppressor allele would change the probability of a phenotypical Suri offspring to only 50 percent.

In this case, the dam, a "pure" Huacaya, has one suppressor allele.

		Suri sire	
		Sp	Sp
Huacaya dam	sP	Ss Pp	Ss Pp
	sp	Ss pp	Ss pp

The genetic contributions from the parents at these two loci sorted themselves out to:

50% **Ss Pp** = Genetic Suri/Huacaya phenotype
50% **Ss pp** = Genetic Suri/Suri phenotype

In other words, <u>all</u> offspring are, genetically speaking, Suris — only half can't express the Suri fiber phenotype due to the dominant suppressor allele.

The <u>dominant</u> allele (**P**) at one locus changes the phenotype of an animal's fleece to mimic that normally created by the combination of two <u>recessive</u> alleles (**ss**) at another locus.

The genotypes **SS PP, SS Pp, Ss PP, Ss Pp, ss PP, ss Pp**, and **ss pp** will produce identical phenotypes — all these alpacas will look like Huacayas. Of course, the suppressor allele will have no effect on the **ss** animals one way or the other. Technically speaking, it is "expressed" — it just doesn't have the opportunity to suppress anything.

Remember when I told you in a previous chapter that the dominant allele is always expressed? This is true, as long as no <u>genetic action at another locus</u> impact its expression. You just learned how easily that could happen.

Although the suppressor allele may be recessive, Sponenberg tells us that breeding results fit more "with a dominant mode for Suri suppression than with a recessive mode." Interested alpaca breeders would do well to read the chapter in Hoffman's book in its entirety.

The important message here is that two separate loci are involved. This leads us rather smoothly to the concept of polygenic inheritance, which is our next chapter.

Chapter Thirteen

POLYGENIC INHERITANCE

CHAPTER THIRTEEN

Polygenic Inheritance

"Polygenic: pertaining to or determined by several different genes."
Blood & Studdert, *Baillière's Comprehensive Veterinary Dictionary*, 1988

Not all traits are inherited in an orderly, easily recognized Mendelian fashion, nor do all genes have only two allelic choices. Height, for example, is not passed on as a simple autosomal dominant or recessive trait. Head shape, length of tail, angulation, length of back, coat- or fleece density and quality — characteristics that can be so important to breeders — cannot be efficiently categorized and assigned to one specific locus or gene.

Most physical traits are determined by a group of genes. Mendelian principles still apply, but the many *polygenes* involved complicate identification and thwart our comprehension. We no longer enjoy the simplicity of dealing with two genes at only one locus coding for a specific trait. The prefix "poly" translates into "more than one," lending the literal definition of "more than one gene" to the term *polygenic*. You will also hear geneticists or breeders use the terms *additive gene action* and *quantitative* traits in relation to the polygenes.

The various shades of human skin color serve as a perfect example of a polygenically inherited trait.

While coat- or fleece color inheritance involves more than one pair of genes, the effects of a single gene pair can be much more dramatic in animals than in humans. On the other hand, some mammalian color genes produce only subtle results such as variations in the basic pigment's shade. The combined genetic control of <u>many genes</u> is necessary for animals to express their final color phenotypes.

While the understanding of simple dominant and recessive traits aids the breeder in selecting for color and against many defects, the serious breeder quickly realizes that polygenes play a more important role.

Polygenically inherited traits can be measured on a mathematical scale, such as the differences in height at the withers (top of the shoulders) among members of a specific breed or species.

Let's explore this further.

The height of an animal, as we discussed, is not inherited in classic Mendelian fashion involving only two alleles at one locus. As breeders, we do not have the simple choice of tall *vs.* short (*discontinuous variation*). If you breed the two extremes, they will not, as Mendel's plants did, sort themselves out as **TT**, **Tt**, and **tt** in subsequent generations. Quite to the contrary, you will have all sorts of *continuous variations*. We call traits under such genetic control *quantitative traits*. The genetic action that controls them is called *additive gene action*. You can remember these terms if you picture a breeder striving to increase the size of his animals. While continuously selecting for larger size, he'll slowly but surely <u>add</u> more and more inches to his stock — or, in the reverse, he can <u>subtract</u> inches, making his animals smaller.

Webster's Dictionary includes these definitions for the word *quantity:* "An amount, portion, and the property of anything which can be determined by measurement." There you go! Height can be measured, length of muzzle can be measured, the exact angle of a croup can be determined, length of wool, micron count — the *quantity* of such traits can be mathematically documented, therefore they can be described as *quantitative* traits.

We have already learned that, while Mendelian principles apply in such cases, the number of genes in a series is often so large that they cannot be readily identified.

Quantitative traits enjoy a high factor of *heritability*, meaning they are easily influenced and modified by selection. Unfortunately, that is why it is so easy for some breeders to play their unique version of "Dr. Frankenstein." I don't want to rush ahead, so we'll save that thought for later.

The Sheep Production Handbook (1992) lists ewe fertility as having a 5 to 10 percent heritability factor, while staple length of fleece showed an impressive 55 percent. Plainly put, through selective breeding it is easier to improve fleece staple length in your sheep breeding program than overall fertility.

A BREEDER'S GUIDE TO GENETICS
Relax, It's Not Rocket Science

Many polygenically inherited traits are also greatly influenced by the environment surrounding the animal. Poor nutrition results in stunted growth, and confinement (lack of adequate exercise) can result in slab-sided, lanky individuals. Environment has been shown to definitely play a part in the occurrence of hip dysplasia, the scourge of so many canine breeds.

Experienced breeders will tell you that certain traits inherited under additive gene action are easier to improve than others. Most dog breeders consider an ugly head to be much easier to "fix" (improve) in subsequent generations than poor angulation. The consensus is that it is more difficult to get rid of coarse heads than refined ones. A "straight" front (lack of correct shoulder angulation) is especially difficult to correct once it has crept into your line. Dr. Julie Koenig, a veterinarian and llama breeder, lists bone, fiber quality (micron count), and fiber quantity as enjoying high heritability in llamas. Such traits are thus easiest to improve in a camelid program. A South African study of German Shepherd Dogs showed that selecting for wider chests quickly resulted in desired measurements.

Polygenes play a role in many genetic defects, such as cardiac problems and hip dysplasia. Such defects only express themselves if enough defective genes come together in one animal. We call that *genetic loading*, or reaching a *genetic threshold*. In other words, the gene dosage determines the extent of the defect's *expression*.

Dr. Jerold S. Bell (Tufts University School of Veterinary Medicine) tells us that many polygenic disorders "have a major recessive or dominant trigger gene that must be present to produce an affected individual" (*The Aristocrat*, Borzoi Club of America, 2000). A gene that acts in a *Mendelian* (dominant or recessive) fashion is a *qualitative* gene. The tricky part here, as Dr. Bell points out, is the possibility of an animal carrying many genes coding for a disease but appearing normal because its genome lacks the (qualitative) trigger gene. Consider the impact this could have on your breeding program. Try to get a copy of this article from your friendly Borzoi breeder. It is very informative and well written.

In relation to polygenes, reviewing the concept of gene linkage is important. Genes carried on the same chromosome are described as "linked." Thus, two unrelated traits may be inherited together due to their linkage. Examples are purple eyes and vestigial wings in the fruit fly, that pesky insect made famous for its contribution to science. Other linkage groups are red eyes and normal length wings.

Researchers have linked specific coat colors to the extent of flight zones in various species. In cattle, hair whorls are reportedly linked to temperament. Horse breeders often link coat color and behavior. I used to think of all this as hokey nonsense. Conclusive evidence that pigment or lack of it influences neurological functions gives credence to the "nonscientific" observations made by breeders in the past. The researcher Edward O. Price and a fellow scientist confirmed a genetic linkage between the recessive non-agouti allele coding for black coat color and temperament in Norway rats — a hypothesis established decades earlier by another scientific worker.

With research spanning five decades, all scientists involved in the studies found the homozygously black Norway rats easier to handle than heterozygous animals or those homozygous for the agouti allele.

Other researchers found that rabbits, bred in a German laboratory for high resistance to infection, started to bite off their own toes in a classic case of self-mutilation. A Russian study of wild fox showed that "continuous selection for a calm temperament in foxes resulted in negative effects on maternal behavior and neurological problems" (Temple Grandin, et al.). This statement should not be interpreted as an endorsement for flighty and nervous temperament. It serves only to explain how care must be taken when breeders apply selection pressure. Temple Grandin, et al, who shared this information in *Genetics and the Behavior of Domestic Animals* (1998) wrote: "Over selection for a single trait ruins the animals. In breeding animals, people must be aware of the complex interaction between traits that do not appear to be related." The message is: <u>Do not appear to be, but are</u>! If traits are linked, selecting against one will perhaps also cause the unintended disappearance of the "mate" connected to it.

Although one or several genes can determine one trait, there are cases where one single gene affects two seemingly unrelated traits. White tigers that are also cross-eyed serve as a good example to explain this genetic action called *pleiotropy*. Another good example is Grey Collie Syndrome, otherwise known as *Canine Cyclic Neutropenia*.

A certain grey in the Collie breed is inherited along with this blood disease, preventing the animals from developing an immune system. The recessively carried gene not only has a deleterious effect on the blood, but "also has been instrumental in causing some kind of interference with melanin formation" (Willis, 1989). It is always lethal. Collie breeders therefore select against this color for health reasons, not aesthetic ones.

There are many such examples. Novices are seldom aware of them, and should investigate their sources before making rash statements accusing experienced breeders of applying irrational and purely cosmetic selection pressures.

Ian Wilmut tells us in *The Second Creation* (2000): "Pleiotropy severely limits the scope and power of genetic engineering, at least given the present state of knowledge." Geneticists want the transferred gene to express only the trait they choose, therefore "they must focus on the minority of characters that have a simple genetic basis, and that are determined by genes that are not highly pleiotropic."

This information came as a surprise to me. A not-so-very old book had presented pleiotropy as an exception to the rule. Since its publication, scientists have obviously discovered otherwise.

Once imbedded in our brains, beliefs based on erroneous and obsolete information are sometimes hard to shake loose. The next chapter discusses breeders clinging to such "truths" like the mouths of nursing puppies to the teats of their dam. If you have ever tried to dislodge one of those little suction cups, you know what I'm talking about.

Chapter Fourteen

FACT OR FICTION?

CHAPTER FOURTEEN

Fact or Fiction?

Ingrid: *"Everybody knows a repeat breeding is never as good as the first. Just look at the four of us."*
Ingrid's three younger siblings: *"You have nerve!"*

Over the years, breeders have put their faith in a potpourri of beliefs strange enough to make a geneticist weep tears of despair. Dog breeders believed for many years that each stud used on a bitch left something of his "essence" in her "blood," thereby influencing future litters. Many bitches of outstanding conformation and temperament were discarded from breeding programs after accidental matings with males deemed undesirable by their breeders.

Special "marks" found on children or animals were explained by the mother being exposed during pregnancy to something scary, or something resembling the blemish, defect, or unusual coat patterns and markings.

In the past, some scientists and breeders believed in the idea of acquired performance traits, a most ridiculous assumption. We can call this the *Lysenko Theory*. Lysenko, a scientist in the Soviet Union, managed to convince government leaders of this theory as it pertains to agriculture, and caused untold damage in that sector. (The originator of this belief was actually a French biologist named Jean-Baptiste Lamarck, who preceded Trofim Lysenko by more than a century.)

Lysenko's preposterous theory claims that, for instance, only actual results achieved by a dog during hunting, racing or retrieving events — not genetic potential — will have an impact on future progeny. We know that such a theory is not based on any scientific facts. Some breeders believed in the inheritance of acquired physical traits. It's all nonsense, of course. As veterinarian Dr. Nina Beyer explains, "If acquired traits could be inherited, puppies born to parents with cropped ears or docked tails would then be born with ears already cropped and tails already docked!"

Dr. Beyer is correct, of course. A super-fast Whippet that is severely injured during its first formal race meet and never raced again will certainly retain the ability to pass on its racing prowess. Its genetic potential will not be diminished in any way by the injury.

Even now, breeders who should know better seem to shy away from using superior individuals that, for whatever reasons, never had a chance to "show their stuff" long enough to make it into the record books. We will explore this concept in more detail in the chapter covering nurturing and nature.

In *Genetics of the Dog*, Malcolm B. Willis writes about a genetic theory proposed by the Australian breeder Bruce Lowe. The top and bottom lines of a pedigree, running from sire to his sire to his sire etc. and from dam to dam to dam, are called the *tail male* and *tail female* lines, respectively. Lowe claimed that the animals represented in those two tail lines influence the genetic make-up of the progeny more than the others. I cannot imagine any person possessing an ounce of logic believing this theory. Neither did Dr. Willis. He explains that if the tail ends of a pedigree do influence a breeding result to an enormous degree, it's because these particular animals were or are what we call "prepotent" (homozygous at many loci), not because of their place on the pedigree. Makes sense to me!

Another Bruce Lowe theory, and one discussed by Dr. Willis, is "let the sire of the dam be the sire of the dam's dam." I've heard breeders repeat this and have seen authors quote it as if it was a magical formula. It does roll so smoothly off the tongue that one is tempted to simply accept it as breeding wisdom — to be followed without the benefit of any scientific, logical explanation.

Before we continue any further, Denise suggested that you should know how to read and understand a *pedigree*, or ancestral record. Pedigrees can be in chart form, bracket form, table form — any form that makes sense and is easy to follow. However, the same format is always present. For each pair, the sire is on the top line, the dam is on the bottom line. In straight-line form, the sire is always mentioned first, the dam second, usually separated by an "x" (Handsome Dan x Pretty Molly). The terms "*by*" refers to the sire, "*out of*" refers to the dam — so Sparky would be "by" Handsome Dan, "out of" Pretty Molly. *Paternal* refers to the sire's side, *maternal* to the dam's side.

A BREEDER'S GUIDE TO GENETICS
Relax, It's Not Rocket Science

The records can continue *ad infinitum*, but the progression of generations is always presented in the same fashion — the following sample pedigree shows 4 generations.

Starting from the left, the progression is:

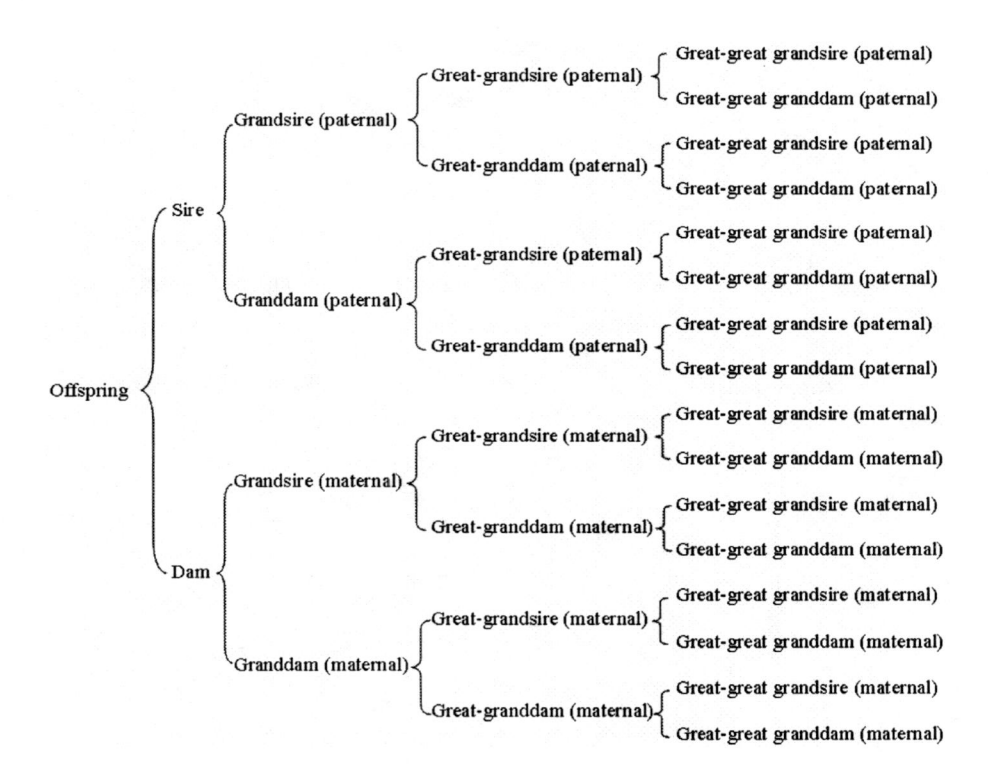

Offspring
- Sire
  - Grandsire (paternal)
    - Great-grandsire (paternal)
      - Great-great grandsire (paternal)
      - Great-great granddam (paternal)
    - Great-granddam (paternal)
      - Great-great grandsire (paternal)
      - Great-great granddam (paternal)
  - Granddam (paternal)
    - Great-grandsire (paternal)
      - Great-great grandsire (paternal)
      - Great-great granddam (paternal)
    - Great-granddam (paternal)
      - Great-great grandsire (paternal)
      - Great-great granddam (paternal)
- Dam
  - Grandsire (maternal)
    - Great-grandsire (maternal)
      - Great-great grandsire (maternal)
      - Great-great granddam (maternal)
    - Great-granddam (maternal)
      - Great-great grandsire (maternal)
      - Great-great granddam (maternal)
  - Granddam (maternal)
    - Great-grandsire (maternal)
      - Great-great grandsire (maternal)
      - Great-great granddam (maternal)
    - Great-granddam (maternal)
      - Great-great grandsire (maternal)
      - Great-great granddam (maternal)

According to Dr. Willis, Lowe's proposed program tried to exploit the fact that "certain sires produced very outstanding daughters rather than sons." With such a pedigree, Dr. Willis points out, a bitch could possibly carry both **X** chromosomes coming from this one sire.

My interpretation of such a pedigree is as follows:

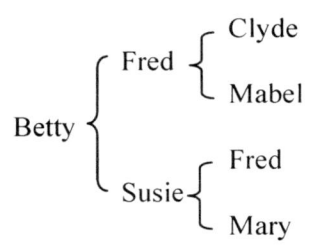

```
                    ┌─ Clyde
           ┌─ Fred ─┤
           │        └─ Mabel
   Betty ──┤
           │        ┌─ Fred
           └─ Susie ┤
                    └─ Mary
```

Let's imagine that Betty is the dam of the prospective litter. To keep it simple, I left out the stud's side of the pedigree. The <u>dam</u> is Betty. The <u>sire of the dam</u> is Fred. The <u>dam's dam</u> is Susie. The <u>sire of the dam's dam</u> is Fred.

Translated into **X** and **Y** chromosomes, in chart form it looks like this:

		Clyde
	$X^A Y$ Fred	
		Mabel
$X^A X^A$ Betty		
		$X^A Y$ Fred
	$X^A X^B$ Susie	
		$X^B X^C$ Mary

Here is the fly (gene?) in the ointment (actually, several flies, as I see it): you can see from the pedigree that Susie could have just as well passed on X^B, making Betty $X^A X^B$, thus reducing Fred's impact on her genome. In addition, according to Dr. Willis, the X chromosome suffers from relative genetic poverty in comparison to the autosomes (a chromosome other than a sex chromosome).

We can speculate that Fred's other desirable chromosomes might also be passed on to Betty as "doubles." We know that chromosomes are inherited randomly, so while it is possible that such desirable doubling up occurs, it is also possible that it does not.

<u>Any</u> linebreeding (which this pedigree represents) has the potential to make the genetic material in the linebred individual more homozygous (more on this later). That said, I cannot see how the exact placement of the superstar's genes in the pedigree would make a difference. The originator of this theory cannot be blamed for the fact that some breeders harbor the erroneous belief that blindly following such a formula will produce superior offspring. Nothing could be further from the truth. The male used for such a program better be as near-perfect a specimen as a breeder can find.

"Never do a repeat breeding. It never produces the quality of the first." Which breeder hasn't heard this odd belief expressed? (I tried to convince my younger siblings for years of the validity of that statement. They ain't buyin' it.) Of course, a repeat breeding will not give you an exact replica of the first result. To conclude from this that germ cells are somehow programmed to automatically carry inferior genes the second time around — as if the reproductive system has a memory of the first mating — is almost too silly to discuss.

Maybe some breeders view germ cells as belonging to a genetic football- or basketball team. First you send in your starters, then the second string — and finally the scrubs? How and where does such nonsense begin to circulate? How does its telling survive for so many years? Who knows?

I am not sure why so many breeders believe in the "fact" that the male always matters more in breeding than the female. Their theory is that a good male can "fix" whatever faults the female brings to the mating. While that is sometimes true, it is no more and no less likely that an outstanding female will "fix" conformational and other faults coming from the male.

Because of the female's exclusive contribution of mitochondrial DNA and other goodies in the cytoplasm, the egg cell's genetic contribution to the zygote is <u>larger</u> than the sperm's (although at the present time mitochondrial DNA is not believed to directly affect an animal's appearance).

When <u>is</u> the selection of a male more important?

In commercial agricultural operations, the bull's genes, for example, <u>are</u> significantly more important than those of the cows he services. That's because frozen or chilled sperm from <u>one</u> bull is routinely used on <u>hundreds</u> of females. His genes will be more prevalent in a population than those of any single cow. However, there is no comparison between such huge operations breeding for meat consumption and hobbyists breeding small colonies of companion animals.

In my opinion, A.I. (Artificial Insemination) should not be an option for any species or breed not specifically bred for slaughter. The danger of genetic drift — a severe loss of diversity — is too great. Human greed and ego will win out over other considerations, especially where large amounts of money are at stake. Ethical breeders and worse — the animals themselves — pay the price. While A.I. is sometimes used as a matter of convenience (to avoid long travel or shipping of a bitch, for example), the potential for genetic devastation of an entire breed population looms ever large. A showring winner's sperm can be used on hundreds of females before a particular genetic defect makes its appearance. True, this can happen with natural breedings as well, but the odds are greater with A.I. that certain males are used on disproportionately large numbers of females.

In that context, we need to examine the use of the word "herdsire" employed by llama- and alpaca breeders. Its choice implies that one male is used to impregnate an entire herd. Some breeders of both small as well as large herds are undoubtably doing that. With only a tiny fraction of North American camelids being slaughtered and consumed, we might question the genetic wisdom of using "herdsires." Maybe the use of "studs," individually selected to correct the faults and enhance the virtues of individual females, might be a more appropriate use of the males in the industry.

Some breeders may envision llamas and alpacas becoming as numerous as sheep. In that case, the use of herdsires versus individual studs will become a financial necessity, unless camelid breeders are willing, as most serious dog breeders are, to take a financial loss with each mating.

To sum it up: a male's genetic contribution is only more important if he produces more offspring in his lifetime than any single female in your kennel or on your pastures.

Participants on the Internet's Alpaca Site joked about hanging pink water buckets in their barns to encourage the conception of female crias. In the "old days," East Prussian farmers paid special attention to the sex of the first visitor entering their barn after the celebration of the New Year. A male visitor supposedly ensured the birth of female animals for the following year. After the birth of several males, you might consider placing a statue of St. Anthony in your barn (just kidding!). Oh, well, whatever floats your boat — it is harmless enough entertainment. There is nothing wrong with having a little fun with your animals, or even much fun, for that matter.

Part of that fun is studying, exploring, and finally deciding on a breeding program — the path you choose to improve your stock. In the following chapters, we will discuss a variety of such programs.

Chapter Fifteen

OPPOSITES ATTRACT — SOMETIMES

CHAPTER FIFTEEN

Opposites Attract — Sometimes

Assortative mating: sexual reproduction in which the pairing of male and female is not random, but involves a tendency for males of a particular kind to breed with females of a particular kind.
Robert C. King, *A Dictionary of Genetics*, 1974

"Well, when we met, I was sitting in the back seat of a car. By the time I stood up, it was too late. I already had him hooked." We laughed as the four-foot-ten-inch tiny and talkative Josette regaled us with stories about her first date with the six-foot-plus tall and quiet Allen, her husband of twenty-five years. Would they be married if she had greeted him "standing short"? Probably. There is plenty of evidence that the folksy phrase "opposites attract" was coined for a reason. Conversely, we see couples that resemble each other enough to look like siblings.

When we select mates for our animals, we may choose to disregard pedigrees to a certain extent and primarily base our decisions on phenotype alone. There are two ways to do this.

One way is called *like to like* (assortative) mating. The other, logically enough, is called *unlike to unlike* (disassortative) mating. Let's explore both concepts.

When breeders refer to like-to-like mating, they usually understand it to be the practice of selecting a breeding pair that closely resembles each other. The pair would be uniform in type, conformation, size, and possibly color. Hopefully, both the sire and dam would exhibit excellent qualities. Remember that animals similar in appearance are not necessarily similar in genetic make-up. Selecting for just one similar trait, for example the shape of the head, is a modified version of the larger concept.

A good rule of thumb is that polygenic traits are more likely to increase quickly in homozygosity under such a program than more simply inherited traits such as coat color (although novices usually believe the opposite to be true).

Like-to-like mating is viable and useful when pedigrees (as in the camelid industry at present) encompass only two or three generations, and the relationship of individual animals to each other is largely unknown.

Don't become rattled when authors tell you that like-to-like mating increases variations of phenotype. You may ask, "Isn't that a complete contradiction? How can breeding animals that are alike increase variety?"

The above statement makes us consider our tiny starter-herds of llamas or sheep, our one or two matings a year. We think of the occasional breeding we do every few years when we want to retain that one special puppy to have a new generation to show or to race. Applied to such a tableau, the idea of variation does not become immediately apparent in our kennel or pasture.

Now picture a breeder with 50 llamas trying to develop two distinct strains — one group for pack animals, the other for fiber production. Over the years, he evaluates each new cria. He assigns it to one group or the other, depending on whether its phenotype leans more toward a large physical frame or an abundance of fiber. In effect, this breeder is mating like to like and thus creating two separate herds.

It takes a long time — but eventually two distinct types evolve and breed true. Isn't this what happened in the llama industry, and with many other breeds or species as well?

I hope this clarifies what appears to be a contradiction to you. I well remember my own confusion when encountering that statement for the first time. As a small breeder, you don't always immediately "think big." Instead of saying that like-to-like matings increase variation, it might be more useful to state that such a program <u>organizes the already existing potential varieties into more homozygously pure groups</u>.

Dog breeders usually do not think in such terms, since their stock is expected to conform to a specific standard for their chosen breed. Genetically speaking, that philosophy has its own pitfalls.

There exists a good example of like-to-like matings erecting genetic walls, and dividing dogs and their owners into two firmly entrenched camps with two separate registries. I asked Denise to explain it. She has more knowledge than I do of the chasm between racing Greyhounds and "show"

Greyhounds, where like-to-like breeding has produced what are now essentially two separate types within the same breed.

"Racing dogs are bred for one reason, and one reason only: for speed. Secondarily, they are chosen for overall health and soundness. Unthrifty, malformed, over-sized Greyhounds, or Greyhounds with extreme front and rear angulation, cannot race successfully — therefore by their very nature they are not used in any breeding programs. The idea is to mate the fastest and the best. As an example, if you look at issues of the *Greyhound Review* (the official racing Greyhound publication in America), you will notice conformity in type and body build — items like coat color, ear set, head shape, and eye color do not matter. Performance matters. After selectively breeding for speed and performance, a basic body type evolves — lean, muscled, moderate in build and angulation, not extreme in any way. Extreme 'parts' result in loss of speed and maneuverability. The 'show' Greyhounds are usually larger, taller, heavier, have deeper chests than necessary (so-called 'keel briskets'), and more rear angulation than they can possibly use. The result? Greyhounds that can no longer function as athletes."

In contrast, unlike-to-unlike mating can be thought of as corrective mating. Unless a breeder is barn-or kennel-blind, he will see faults in his animals he would like to correct and features he would like to improve on in future generations. They are not always glaring problems, but can be minor nuisance faults such as ringed tails or light eye color that are not acceptable in certain breed standards.

Do not always think in terms of the sire correcting the faults of the dam. It works just as well the other way. A few years ago, I bred my foundation bitch to a stud offering many traits I admired and whose pedigree meshed nicely with that of my bitch. Unfortunately, he had a slight overbite. Although his bite was not enough to disqualify him from performance events or the show ring, he was not even considered as a potential stud by other breeders. My bitch, her sire, her dam, her littermates, and the stud's parents and siblings all had excellent bites. I also knew that this dog and his siblings had been neglected as tiny babies — they were rescued and nursed back to health by the sire's owner. This led me to speculate that environmental conditions might have played a part in developing his poor bite (admittedly a long shot, but a consideration nonetheless).

Based on these observations, I took a chance and did not regret it. The breeding produced four puppies, all with excellent bites. The bitch I retained from this litter later had eight offspring. All eight have perfect bites. To be sure, not all such gambles pay off, and risking them is foolish if the fault is severe or the animal is not outstanding in other respects. Also, be aware that certain faults can crop up again in subsequent generations. Jaw length, by the way, is inherited independently for the upper and lower jaws.

A common misconception concerning unlike-to-unlike matings unfortunately still circulates among breeders. The author of an article discussing various breeding programs suggested breeding a sickle-hocked alpaca to a post-legged stud as a corrective measure. (Denise was horrified to learn that this fallacy is still believed, and worse, acted on.) There are those who feel that anyone is entitled to voice his or her opinion, and breeding decisions are simply a matter of preference. They are absolutely correct. Anyone who wants to compound one major fault by breeding to another is certainly entitled to do that. Let us hope the animals will not have to pay the price. Poor angulation, for instance, usually reaches beyond the cosmetic and eventually causes orthopedic problems.

Other breeders might prefer to follow advice that has stood the test of time. Knowledgeable breeders would say: <u>never, ever try to correct a fault by breeding to the fault at the opposite end of the spectrum</u>. A post-legged or sickle-hocked alpaca should be bred to one with <u>correct</u> angulation (or, as Denise said, not bred at all). An Afghan Hound with an overbite should never be bred to one whose bite is undershot. It should be bred (if at all) to an animal with a <u>correct</u> bite and come from a long line of animals with correct bites. A Borzoi with a wheelback (roach) should not have a mate selected whose topline reminds one of a hammock. A Whippet bitch with a short neck would be ill-served by mating her with "Mr. Giraffe." The breeder of a base-wide female llama should look past the stud that has "both legs coming out of the same hole."

Please do not contemplate using the alpaca example as a method to correct temperament, by breeding a shy animal to an aggressive one. It does not work! There are enough Dr. Jekylls & Mr. Hydes in all species — and some of them are dangerous to other animals and to humans.

Let me preface the next few chapters by saying that I feel that there is a place for more than one breeding program. I do not advocate a single concept above all others. Read the information, discover what others say on the subject, and talk to experienced breeders who have produced the type of animal you prefer. Then decide which method suits your particular situation and goals. Beware of authors who present overly simplified solutions to complex problems. All breeding programs carry the potential for positive results as well as negative repercussions. There is no "one size fits all."

Let's confront an issue resulting in heated arguments among breeders of all species. Don't worry, no shots are fired, but people do get steamed up over this subject.

Chapter Sixteen

INBREEDING

CHAPTER SIXTEEN

Inbreeding

"Inbreeding is the mating of related individuals (usually not more than two generations removed from one another), where neither individual is an ancestor of the other."
Journal of Canine Genetics, Society for the Advancement of Canine Genetics, (quoted from the *Irish Wolfhound Guide*, Alfred W. DeQuoy, 1973)

Most cultures carry a strong taboo against what is commonly called "inbreeding." It is against the law in many countries. We call it incest when humans practice it. Most of us can hardly imagine something so vile happening to us or freely participating in such an activity. I still remember my open-mouthed horror when first hearing about "it" as an (admittedly naive) teenager. Later I read articles about the various European royal families, all genetically intertwined through politically motivated marriages. They often produced hemophiliacs, the result of a gene carried as a sex-linked recessive by women. Passed on to their sons on their X chromosome, the boys' Y chromosome cannot offer a healthy allele to countermand X's genetic orders.

Scientists identified Queen Victoria of England (1819-1901) as the original source of the royal problem. The unfortunate gene even caused political and historical repercussions. The Russian Tsarina fell under the spell of the evil monk Rasputin while trying to save her hemophiliac son. This enraged the populace, contributing to the Russian Revolution and the subsequent murder of the Czarist family.

Since none of Queen Victoria's male ancestors displayed the trait, there is speculation among geneticists that the Duke of Kent was not Queen Victoria's biological father (no hate mail from British subjects, please!). If Queen Victoria's "mum" didn't cuckold her husband, the cause of Victoria's DNA coding for hemophilia might have been a *spontaneous mutation*.

Because so many of us anthropomorphize our animals (I plead guilty as charged), we subconsciously carry the taboo of incest over to our breeding programs. When Piggybank, my three-month-old Whippet male, mounted his sibling in play, her owner cried in mock terror, "Get off my little girl, you pervert. That's your sister!"

Breeders can't agree on what actually constitutes inbreeding versus the more universally accepted linebreeding. The standard joke is that if a breeding produces great results, you call it linebreeding. If it produces garbage, you call it inbreeding.

Reading the conflicting definitions of several authors is enough to drive one crazy. Alfred W. DeQuoy, the well-known Irish Wolfhound breeder, quotes the Society for the Advancement of Canine Genetics. An article published in 1947 in the *Journal of Canine Genetics* states: "Inbreeding is the mating of related individuals, where neither individual is an ancestor of the other." Brother to sister would be inbreeding, son to dam would not. Since the dam is her son's ancestor, the Society would classify such a mating as a linebreeding.

Now we go to Dr. Willis, who states: "The correct definition of inbreeding is the mating together of animals more closely related to one another than the average relationship within the population (or breed)." He then explains the evil-looking formula for figuring out the *co-efficient of inbreeding*. Don't worry, you will learn to calculate this in another chapter. Van Vleck explains in *The Horse* that when some ancestors "are related, inbreeding results, and the problem of determining relationships is more complex." I could go on and on.

What are we to do? Breeders usually differentiate between *inbreeding* and *linebreeding*. This only makes sense to me if the two systems have different objectives. If it's just a matter of degree as so many breeders claim (inbreeding is often vaguely defined as a more "intense" form of linebreeding), why bother to use two different labels?

Many years ago, when I read the definition of inbreeding as quoted by Alfred W. DeQuoy in *The Irish Wolfhound Guide* (1973), I actually thought it included a typographical error. How in the world, I wondered, can a mother-son or father-daughter mating not be defined as inbreeding? It took much more study before I finally understood how a brother-sister mating (the most intense form of inbreeding) differs <u>in purpose</u> from a sire-daughter mating (linebreeding).

Both inbreeding and linebreeding result in an increase of homozygous loci in the offspring. So, what's the difference?

Folk wisdom holds that inbreeding <u>in itself</u> promotes general uniformity in a herd, flock, or kennel. It most certainly does not! People erroneously envision such a program as a primitive sort of forerunner to cloning, with *all* progeny closely resembling each other.

Let's examine this concept with a practical example, using a fictitious brother-sister mating. I will assign them letters for genetic traits and their respective alleles.

H = Curly hair
h = Straight hair
B = Black coat
b = Liver coat
C = Short corkscrew tail
c = Long straight tail
D = No defect
d = Defect

For the sake of our discussion, let us say the brother is **HH Bb Cc Dd**. His sister carries identical allelic combinations, except she is **hh**. Therefore, both have black coats, corkscrew tails, and carry — but do not express — a specific defect (let's say albinism). The only difference is that the brother has curly hair, while the sister's is straight. Return to Chapter Four for a refresher on use of the Punnett Square, and then work out possible combinations for the various traits. Is the lightbulb going off in your head?

If the mating of these siblings produced a litter of three, their genotypes and phenotypes could <u>possibly</u> look like this:

Progeny I = **Hh Bb Cc Dd** = Curly hair, black, corkscrew tail, no defect visible (carrier)
Progeny II = **Hh BB CC DD** = Curly hair, black, corkscrew tail, no defect present
Progeny III = **Hh bb cc dd** = Curly hair, liver, straight tail, defect expressed (albino)

You've just gone from 60 to 100 watts, right? The three offspring need not be uniform at all, except one trait where the only choice was **Hh**. Progeny III, although technically liver-colored, is actually white because of the two recessive alleles coding for albinism.

If your aim was to rid your line of the dreaded defect (though you won't know that unless a DNA test is available), and you want your animals to be homozygous for black coats and short corkscrew tails, you've accomplished that goal with Progeny II. You have muddied the waters a bit with coat variety, but three out of the four desired traits isn't bad. You would certainly discard Progeny III from your breeding program, and would do likewise with Progeny I if you could "see" its full genotype. Please remember that the scenario discussed above is totally fictitious — you can easily replace my chosen traits with those applying to your own species or breed.

Camelid breeders can, just for fun, try a "paper" mating of two Suri siblings. Let's assume they're **Ss Aa Dd** — red Suris that both carry but do not express a defect. Well, did you figure out the probability? Did you sit up in total shock when our hypothetical Suri siblings produced a black Huacaya with a defect?

Where does all this talk of uniformity from inbreeding come from? From inbreeding accompanied by rigid selection pressure (culling) — that's where!

Keep in mind that many breedings and much culling must take place to accomplish uniformity and consistency in a large population.

By the way, self-fertilization in various species presents the ultimate inbreeding. Srb, et al, shows a table with 1600 individuals starting out as Aa. Only four generations of *self-fertilization* later, <u>750 progeny carry the **AA** genotype and 750 carry **aa**</u>. Only 100 individuals will still be **Aa**!

Of course, any inbreeding program using mammals will not produce the quick results that self-fertilization does. The concept and the effect, however, are the same. When breeders talk about inbreeding producing uniformity, they often don't understand the correct meaning of that term as it applies to such a breeding program.

Srb, et al, puts it plain and simple: "Inbreeding results in homozygosis or, if you will, the homozygous state at numerous genetic loci." That "state" can be either **AA** or **aa**, to use one particular locus as an example. Ah, don't we all enjoy such knowledge-defining "lightbulb" moments?

That said, we must remember that it is possible for siblings or cousins to be genetically very uniform (homozygous at many loci).

Breeders must be discerning observers of breeding results and know their pedigrees. Don't just assume that "close" matings automatically result in uniform offspring.

This is a good time for a refresher on chromosome activity in the cell (Chapter Three).

The following pedigrees may further clarify the definition of inbreeding as used by the Society for the Advancement of Canine Genetics.

I. <u>Brother-Sister</u> (inbreeding)

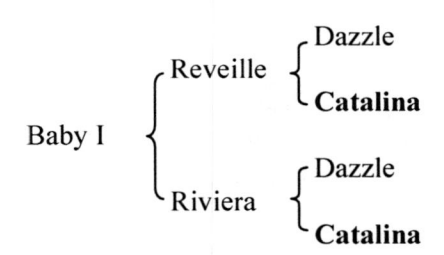

```
                     ┌ Dazzle
           ┌ Reveille ┤
           │         └ Catalina
  Baby I  ─┤
           │         ┌ Dazzle
           └ Riviera ┤
                     └ Catalina
```

II. <u>Sire-Daughter</u> (linebreeding)

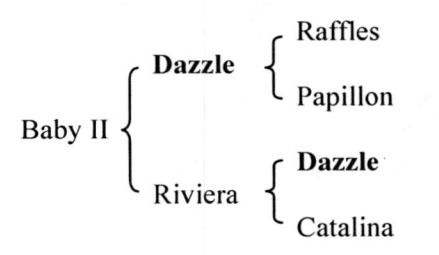

```
                     ┌ Raffles
           ┌ Dazzle  ┤
           │         └ Papillon
  Baby II ─┤
           │         ┌ Dazzle
           └ Riviera ┤
                     └ Catalina
```

In pedigree "I" (inbreeding), it is possible for either Dazzle's or Catalina's genes to be "shut out" completely from Baby's genome. Pedigree "II" (linebreeding) assures that Dazzle's genes will be passed on to Baby. Baby II is linebred "on Dazzle."

III. <u>Son-Dam</u> (linebreeding)

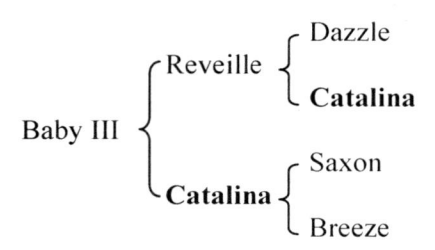

```
              ┌ Reveille ┌ Dazzle
              │          └ Catalina
    Baby III ─┤
              │          ┌ Saxon
              └ Catalina ┤
                         └ Breeze
```

You can also linebreed "on the dam." Baby III is linebred "on Catalina."

Once again, with emphasis: <u>Inbreeding without severe selection pressure is an exercise in futility</u>. It serves no purpose. Inbreeding <u>with selection</u> creates the desired homozygosity in your kennel or barn.

Conversely, the dangers <u>and</u> the benefits of "flushing out" defective genes must be strongly considered. Again, this can produce a very positive outcome <u>if you are willing to cull</u>.

Humans serve as perfect examples of inbreeding <u>without</u> subsequent selection. Steve Jones, a professor of genetics at the Galton Laboratory of University College, London, England, presents fascinating and disturbing examples of the consequences of inbreeding in various small ethnic groups all over the world. I strongly recommend serious study of *The Language of Genes* (1993) in addition to texts on animals before embarking on any inbreeding program. Inbreeding can bring stunningly beautiful results. It also requires a strong stomach and generous amounts of time and money. In a later chapter, I address inbreeding depression and its remedies.

Let us clarify a fine but very important point: Inbreeding does <u>not</u>, as many breeders believe, <u>produce</u> defects. It does not create or generate them the way two sticks rubbed together will generate fire. Defects will only surface if the animals used in such a program carry the recessive genes producing the defect. If a line of llamas, for example, is completely free of *choanal atresia* (strongly suspected to be a genetic defect — although no definite proof presently exists), even a brother-sister mating will not cause this devastating defect to magically appear. The big little word here is "if." Without DNA testing or extensive test breedings, we cannot know which animals carry the genes coding for defects.

Astute breeders of many species use their extensive knowledge of pedigrees — and the individuals represented by them — to utilize inbreeding as the cornerstone of successful breeding programs. The advertisement section of a lure coursing publication recently featured the pedigree of a successful Basenji. Not only had this young male garnered points in the show ring, he also made the prestigious Top Ten list in two separate lure coursing programs. Who was his sire? He was the offspring of a brother-sister mating. So, don't automatically turn up your nose when you hear of a breeder planning such a combination.

Frankly, I never considered the issue of longevity in relation to inbreeding. Perhaps we should! Roe Froman, D.V.M., presented information in the *AKC Gazette* Clumber Spaniel column (October 1999) about a research program she is collaborating on with Dr. John Armstrong of the University of Ottawa. I quote Froman: "The primary objective of this study has been to evaluate the impact of inbreeding on life span and the incidence of genetic problems."

The project, then in its preliminary stages, nevertheless collected sufficient data on Standard Poodles to justify an analysis. In that breed, Dr. Armstrong found that "dogs with an inbreeding co-efficient of more than 30 percent (based upon a 10-generation calculation) die, on average, at about ten years. This is three years earlier than dogs with an inbreeding co-efficient of less than 10 percent." This study deserves to be followed by those interested in using inbreeding in their breeding programs.

Longevity might not be high on the list of important traits to breeders with large turnovers in their kennels. It is important to pet owners, and to small hobby breeders who only put a litter on the ground every five or six years. Likewise, a breeder of meat sheep would not be concerned about longevity in his flock, while a llama- or alpaca breeder would probably consider a long, productive lifespan a desirable trait to breed for.

I've demonstrated that highly inbred offspring can express a drastically different phenotype than either parent. <u>Inbred litters, therefore, will generally show greater variation than linebred litters.</u> I like to think of inbreeding as genetically "sorting out" your stock. (Compare it to unraveling a multicolored sweater and sorting the colors into separate piles.) After you've selected the most desirable individuals, you then proceed to build your breeding program upon the physical and possibly mental attributes of these animals. Many breeders choose linebreeding to accomplish that.

Large programs may eventually carry several "lines," each featuring a genotype and expressing a phenotype similar to those of their *founder*. Such breeders can then *linecross* to animals right in their own kennels or pastures.

Both inbreeding and linebreeding involve the mating of related animals. There are breeders who will agree with me that the philosophy and <u>purpose</u> behind each one can differ significantly. You'll find just as many who will tell you that the difference between the two programs lies only in the degree. Take your pick! I've learned not to argue about this, so if you mail me a note with violent objections to my interpretation of inbreeding, don't expect a response. Adopt the definition that makes you feel comfortable.

Before you decide which "camp" you'll join, let's explore the concept of linebreeding in more depth.

Chapter Seventeen

LINEBREEDING

CHAPTER SEVENTEEN

Linebreeding

"Linebreeding: the type of inbreeding which concentrates on one given ancestor with the goal of concentrating the genetic impact of this individual in the population."
D. Phillip Sponenberg, D.V.M. Ph.D., *A Conservation Breeding Handbook*, 1995

I think linebreeding is a little easier to understand, and breeders feel more comfortable with the concept. It doesn't carry the moral baggage implied by the common use of the word "inbreeding." Linebreeding makes perfect sense when a breeder has the good fortune to either purchase or breed an outstanding specimen. He or she can "double up" on that individual animal.

We haven't visited our fictitious kennel in a while, so let's return and work out some pedigrees. Remember our stud dog, Felix? We consider him as close to the ideal Whippet as we can get. We are thrilled with the progeny produced by his breedings to Ayla and Gladys so we decide to linebreed on Felix in the hope of creating a truly prepotent stud. We therefore breed Jenny, a Felix daughter, back to her sire. This breeding produces Tom. Tom bred to Sunflower results in Max, who exhibits many traits we admire in Felix.

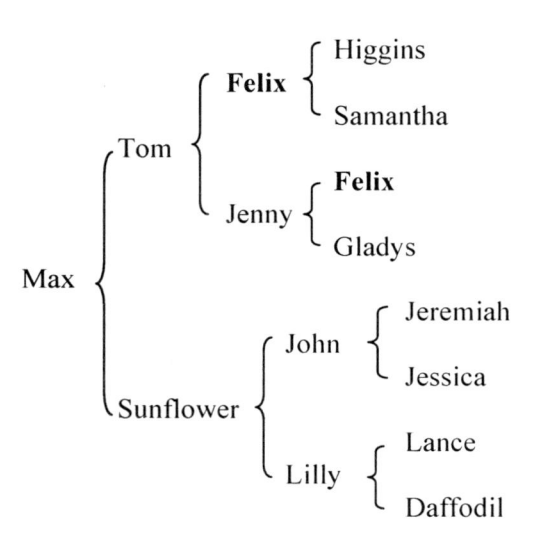

```
                              ┌─ Higgins
                    ┌─ Felix ─┤
                    │         └─ Samantha
            ┌─ Tom ─┤
            │       │         ┌─ Felix
            │       └─ Jenny ─┤
            │                 └─ Gladys
   Max ─────┤
            │                 ┌─ Jeremiah
            │         ┌─ John ┤
            │         │       └─ Jessica
            └Sunflower┤
                      │       ┌─ Lance
                      └─ Lilly┤
                              └─ Daffodil
```

One such breeding does not constitute a "line." This is only the beginning of our planned program. Keep reading.

We decide to further strengthen Felix's genetic impact on our line, and breed his grandson Max to Trudy to achieve that goal. Trudy is another Felix daughter, but is not related to other dogs in Max's pedigree.

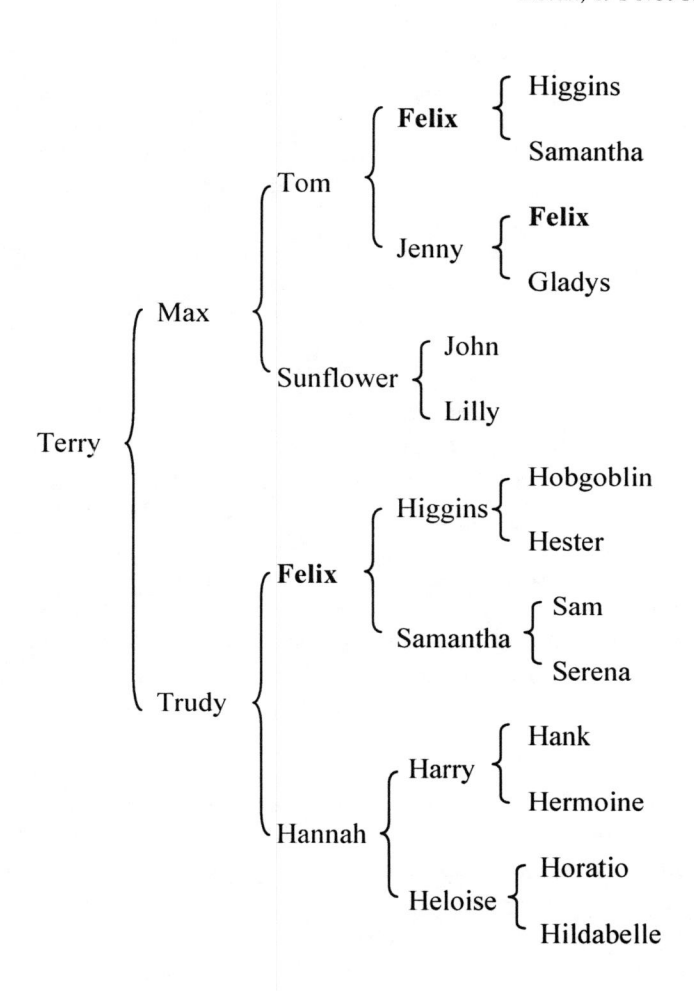

A BREEDER'S GUIDE TO GENETICS
Relax, It's Not Rocket Science

```
Terry ┬ Max ─┬─ Tom ──────┬─ Felix ────┬─ Higgins
      │     │             │            └─ Samantha
      │     │             └─ Jenny ────┬─ Felix
      │     │                          └─ Gladys
      │     └─ Sunflower ─┬─ John
      │                   └─ Lilly
      └ Trudy ┬─ Felix ───┬─ Higgins ──┬─ Hobgoblin
             │            │            └─ Hester
             │            └─ Samantha ─┬─ Sam
             │                         └─ Serena
             └─ Hannah ──┬─ Harry ─────┬─ Hank
                         │             └─ Hermoine
                         └─ Heloise ───┬─ Horatio
                                       └─ Hildabelle
```

Felix appears three times in four generations. If Terry, Felix's grand-son as well as being his great-grandson and great-great-grand-son, "grows out" (matures) to have a beautiful temperament, great conformation, strong performance attributes <u>and</u> an absence of major defects, we have hit the genetic jackpot. Terry has a good chance of being prepotent for the traits we admired in Felix, meaning his genotype is homozygous at many loci. We hope he puts his stamp on the progeny of every bitch he's bred to, siring a string of dogs that carry his likeness.

Since genetics rarely work that smoothly, breeders need to employ common sense and powers of observation when it comes to breeding. If Felix is large, long-legged and has an elegant head, it would be foolish to advertise a small, short-legged, short-muzzled son or grandson as being "linebred on Felix." He is a close Felix descendent on paper, of course, but obviously inherited the bulk of his genes from the "other side" of the pedigree. You can argue that phenotype does not always reflect genotype, which is true. Regardless, as the owner of a bitch (and having known Felix) I would be decidedly unhappy to find such a genetic "imposter" being used to breed my female.

You might hear breeders refer to "loose linebreedings," meaning a common ancestor does not appear until the fourth or fifth generation in the pedigree. Some would describe Terry as inbred rather than linebred. There is no sense quibbling over "labels" with other breeders. No one wins.

Study the two systems — use the Punnett Square, and appreciate the concept of how allelic combinations separate again in future generations. You might agree that the second part of DeQuoy's quote on inbreeding ("where neither individual is an ancestor of the other") is the key to understanding the difference in the two programs. (Okay, so I can't resist quibbling just a bit.)

We can use a single locus to further clarify this difference. With inbreeding, you hope to segregate heterozygous combinations such as **Aa** into homozygous ones — **AA** and **aa** — so you can choose. With linebreeding, you've already made your choice — **AA** or **aa**. You exclusively use the genotype that, due to its homozygosity, breeds true and therefore always pro-duces the phenotype you prefer. Of course, since thousands of loci are involved rather than just the one used in my little sample, it is not an easy task to get all the desired allelic combinations in your breeding program.

Breeders often use the expression "breeding true" (read: parents produce offspring nearly identical to themselves). A linebred animal's *get* (progeny) will quite possibly be as predictable as the art of breeding will allow. Notice I used the word "art." Yes, breeding is a science, but intuition and that hard-to-explain "eye" for a good animal also plays a role. A hefty chunk of luck doesn't hurt either.

Be careful when using the expression *linebred* or *line*. A breeder recently told prospective buyers in a sales brochure that two imported animals had produced a beautiful baby, and she planned to continue "this line." Breeding two unrelated animals, as you have seen, cannot even remotely be described as being the beginning of a "line."

I feel that a young animal should never be chosen for linebreeding. Some defects don't show up until the animal is quite mature. Sometimes promising youngsters "fall apart" (breeders' jargon for getting ugly) once they mature, or they develop strange personality quirks. To invest enormous amounts of time and money developing a line based on an animal that later turns out to be a "dud" is heartbreaking indeed. It is far better to practice patience and restraint! (This sounds suspiciously like the dating advice fathers give to daughters — breeders and daughters alike should listen!)

Breeders of commercial livestock can take more chances, of course. If one of their experiments doesn't produce the desired results, the animals can be taken to market. As an alpaca breeder, I can appreciate the special dilemma this puts me and my fellow breeders in. In South America, alpacas and llamas are considered livestock and inferior animals are eaten. This is not the case in this country, where the cuddly-looking camelids are often treated as pets, complete with clicker-training and red bows tied around their necks at Christmas.

You'd have to possess a heart of stone not to have it melt at the sight of a cria! My husband and I were completely smitten at the first sight of an alpaca baby, who looked just like a big stuffed toy. What brute could possibly think of eating it or its parents? "It's the Disney factor," quipped a former beef cattle farmer now breeding llamas. The last time I looked, his llama babies were all still happily frolicking around, and the old show champion he rescued still enjoys a special comfortable pasture for grazing and chewing his cud.

An alpaca cria is a delight to watch and touch. It should remain with its dam for a minimum of six months to nurse and to be schooled in proper alpaca etiquette.
(photo by Rock Chimney Farm Alpacas)

Not eating the results of poor breeding choices makes breeding decisions more complex. You tend to think longer and harder about the possible ramifications of your actions.

The benefits of inbreeding and linebreeding will eventually fizzle out. As the leader of any even slightly anemic business realizes, sometimes it's necessary to bring in "new blood." Of course, you know by now that it's not blood but DNA that makes the difference. Nevertheless, breeders still refer to *bloodlines*. This term takes on great significance when we endeavor to add different genetic material to our inbred or linebred stock.

Before we explain that concept, let's learn about a mathematical formula called the *co-efficient of inbreeding*. It can be applied to linebreeding as well.

Let's solve some mathematical equations (oh, horrors!). Don't fret, they're not so bad. I'm not ashamed to admit that I make no effort to memorize the formulas. Who cares? Let's not get uptight. Relax, it's not rocket science!

Chapter Eighteen

THE CO-EFFICIENT OF INBREEDING

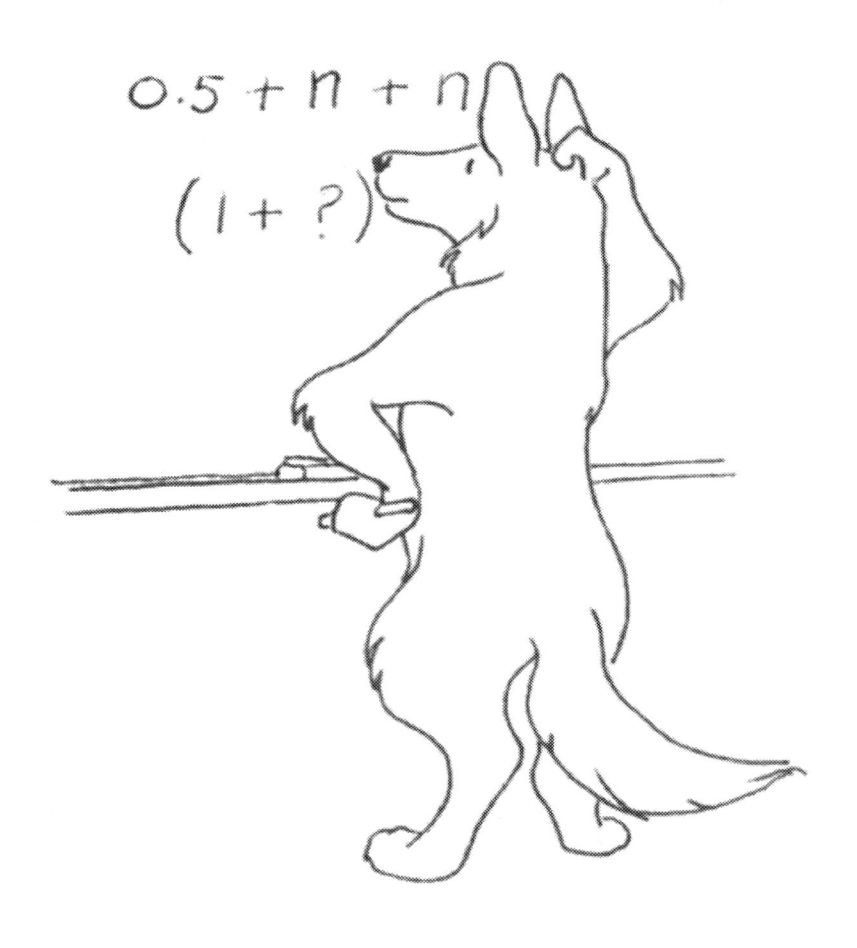

CHAPTER EIGHTEEN

The Co-efficient of Inbreeding

"Co-efficient: 1. An expression of the change of effect producedby the variation in certain factors, or of the ratio between two different quantities"
Blood & Studdert, *Baillière's Comprehensive Veterinary Dictionary*, 1988

In your study of genetics, you'll eventually hear of *Galton's Law* and the *Hardy-Weinberg Law*. You will now read about (tongue-in-cheek) *Wood's Law*. It is very simple. If this chapter leaves you screaming in frustration or curled up in a fetal position admitting defeat, you have no business doing any inbreeding or close linebreeding (my husband hates it when I get bossy like this — sorry, I think this warrants such an attitude). Better to confine your breeding adventures to what geneticists call *like-to-like* mating, using the outcross model. Settle on the phenotype you're happy with, and breed unrelated animals resembling each other to, well, each other.

The formula for the *co-efficient of inbreeding* must be credited to Sewall Wright, who wasn't mentioned in Webster's Dictionary for his herculean efforts (at least not in the 1970 puppy-gnawed edition that I'm still using).

The basic formula reads: $\mathbf{F_x} = \Sigma\ [(0.5)\ n+n^1+1\ (1+\mathbf{F_A})]$

x = co-efficient of inbreeding of X
n = number of generations from x's sire to an ancestor common to the paternal and maternal side. This is zero if the sire himself appears in the pedigree of the dam.
n^1 = the number of generations from x's dam to an ancestor common to the paternal and maternal side. This is zero if the dam herself appears in the pedigree of the sire.
Σ = summation of separate contributions from each different common ancestor.
F_A = co-efficient of inbreeding of the common ancestor, *when that animal is itself inbred.*

Malcolm B. Willis developed a slightly different formula. The results are the same, only the calculations follow a different format. I'll only present Wright's, but encourage you to investigate the other method as well.

Our first three models will not be dealing with F_A, so we can drop that part of the equation for now.

Model I:

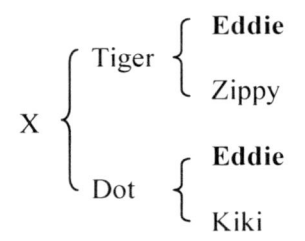

```
        ┌ Tiger ┌ Eddie
        │       └ Zippy
    X  ─┤
        │ Dot   ┌ Eddie
        └       └ Kiki
```

Eddie is the common ancestor. He is not inbred himself (at least this abbreviated pedigree doesn't show any inbreeding), therefore we can drop, as mentioned above, the FA. X's sire is Tiger. We count one generation from Tiger to Eddie, so n = 1. X's dam is Dot. We count one generation from Dot to Eddie, so n^1 = 1. Now that we've figured out the values for n and n^1, let's solve the equation:

$x = (0.5)^{1+1+1} = 0.5^3$
$x = 0.5 \times 0.5 \times 0.5$
$x = 0.125$
$x = 12.5$ percent

Model II:

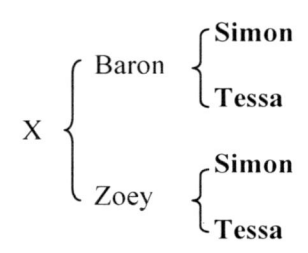

```
        ┌ Baron ┌ Simon
        │       └ Tessa
    X  ─┤
        │ Zoey  ┌ Simon
        └       └ Tessa
```

122

A BREEDER'S GUIDE TO GENETICS
Relax, It's Not Rocket Science

The two sides of the pedigree share two common ancestors (brother-sister mating). In this case, calculate the contributions of Simon and Tessa separately, then add them together. They're both identical (12.5 percent), and add up to 25 percent. Some authors might simply list the combined 25 percent. It is perhaps more logical to describe the offspring of such a mating as 12.5 percent inbred on Tessa and 12.5 percent inbred on Simon.

Model III:

```
                              ┌ John
                    ┌ Rigby  ┤
                    │        └ Josie
           ┌ Sport ┤
           │       │         ┌ Hans
           │       └ Honey  ┤
           │                 └ Helga
    X ────┤
           │                 ┌ Rigby
           │       ┌ Tom    ┤
           │       │         └ Lucy
           └ Sadie ┤
                   │         ┌ Fritz
                   └ Mary   ┤
                             └ Suzy
```

This pedigree represents a linebreeding. "X" is linebred on Rigby. You can see that Rigby was bred to two different females. He is the only ancestor Sport and Sadie share. Neither Sport nor Sadie appear in each other's pedigrees. Please review the formula's description of the calculation. For our model, it works out to this:

n = x's sire to common ancestor = one generation
n' = x's dam to common ancestor = two generations
Rigby is not inbred himself, so once again we can drop F_A.

$X = (0.5)^{1+2+1}$
$X = (0.5)^4$
$X = 0.0625$
$X = 6.25$ percent

Ingrid Wood and Denise Como

Model IV:

This is the real test of your stamina (or your threshold of pain). Send your spouse and children to scoop poop, hose down the kennel walls, brush the llama, halter-train the foal or alpaca — anything to give you enough peace and quiet to concentrate.

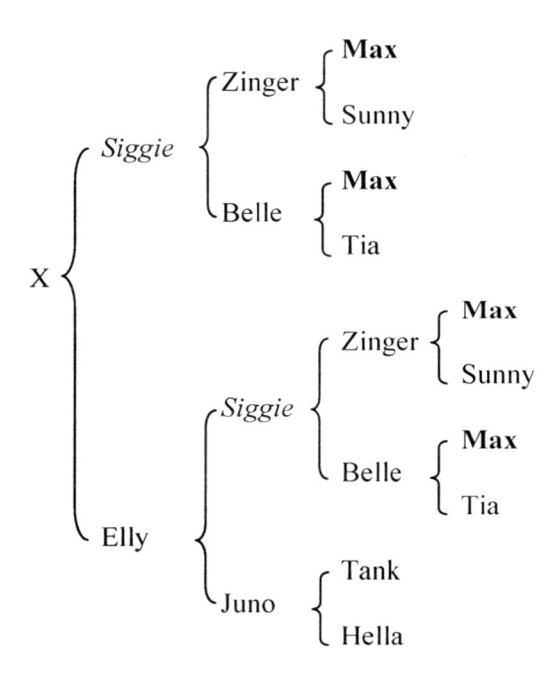

```
                    ┌ Zinger ┬ Max
          ┌ Siggie ─┤        └ Sunny
          │         └ Belle ─┬ Max
          │                  └ Tia
    X ────┤
          │         ┌ Siggie ┬ Zinger ┬ Max
          │         │        │        └ Sunny
          └ Elly ───┤        └ Belle ─┬ Max
                    │                 └ Tia
                    └ Juno ─┬ Tank
                            └ Hella
```

Notice that F_A applies now. Siggie is the common ancestor, and he is inbred himself.

$$F_A = (0.5)^{1+1+1} = (0.5)^3 = 0.125$$

F_A represents the co-efficient of inbreeding for Siggie.

What is the co-efficient for X?

n is zero since the sire himself appears in the pedigree of the dam.

n^1 is 1 — it's one generation from Elly to the common ancestor.

$$X = (0.5)^{0+1+1} (1 + F_A)$$

$$X = (0.5)^2 (1.125) = 0.28125 = 28.125\%$$

Most breeders consider more than 20-30 percent a signal for caution, with 30 percent and higher a dangerous inbreeding co-efficient.

Breeders often don't understand what the percentages represent, so let me explain. What exactly do 6.25 or 25 or 28 percent signify? Adrian M. Srb, et al, tells us in *General Genetics*: "The inbreeding co-efficient measures the extent to which heterozygosity may be expected to be reduced in any individual as a consequence of relationship between his parents." (Emphasis is mine.) Let's use real numbers. If a random-bred animal is heterozygous at 1000 loci, then an individual from the same species or breed with an inbreeding co-efficient of 0.25 (25 percent) is expected to be heterozygous at 750 and homozygous at 250 (25 percent) of these loci.

The author reminds readers that "the absolute genetic value of this measure depends on the extent of diversity in the population concerned." Breeders should therefore view an inbreeding co-efficient as a guideline, not an absolute.

Interested and mathematically inclined readers will find plenty of challenging pedigrees in more sophisticated and advanced texts.

The computer literate among you may want to purchase pedigree software programs to do the job of calculating inbreeding co-efficients for you — which make it very easy to play with hypothetical breeding combinations to see how they look on paper. Several good programs are available, and it really doesn't matter what species you apply them to.

Chapter Nineteen

OUTBREEDING

CHAPTER NINETEEN

Outbreeding

"Outbreeding is a mating between individuals that are less closely related than average. A mating should probably be considered as outbreeding when the two individuals show no common ancestor in a four-generation pedigree."
Journal of Canine Genetics, quoted from *The Irish Wolfhound Guide*, Alfred W. DeQuoy, 1973

Breeders may successfully inbreed or linebreed without their stock carrying or expressing severe genetic defects, or culling them when necessary. However, they must be aware that a phenomenon called *inbreeding depression* will eventually set in. What exactly does this mean? (No, it doesn't mean that the breeder feels depressed.)

As we've discussed, both programs can fix type and produce animals of uniformly superior conformation. However, traits associated with the general vitality of the animals may begin to suffer. Fertility, milk production, overall robust good health, the ability to deal with parasites — all these and more can be (and often are) negatively affected, or *depressed*.

Let us assume that in our fictious kennel all breeding stock is now tightly linebred on Felix. We don't dare push the co-efficient of inbreeding any higher. After careful evaluation, we realize that we are trapped — genetically speaking we have "bred ourselves into a corner." We must *breed out* (or outcross, as some breeders refer to it) to revitalize our own line. We have two choices: we can choose an animal that is linebred itself (linecrossing), or that is a complete outcross (an individual showing no common ancestor in a four-generation pedigree). In either case, we have to find an animal belonging to the same breed as our own, but that is not related to our stock.

This statement is open to interpretation. Breeds often evolve or were created from a very tiny gene pool. Sometimes a famous stud was used on hundreds of females. You will find pedigrees showing the same animal over and over again, going back ten or more generations. At other times, only careful study will possibly reveal genetic relationships you did not recognize at first glance. A two- or three generation pedigree might not show a single name in duplicate, yet may actually result in a high co-efficient of inbreeding if you investigate further.

Only the great-grandparents tell the tale!

```
                    ┌ Anthony ┌ Jim
                    │         └ Claire
          ┌ Tony ───┤
          │         │         ┌ Larry
          │         └ Rita  ──┤
          │                   └ Linda
Bella ────┤
          │         ┌ John  ──┬ Jim
          │         │         └ Claire
          └ Lisa ───┤
                    │         ┌ Jim
                    └ Nancy ──┤
                              └ Claire
```

Without seeing the ancestors furthest to the right, you would have no clue that Anthony, John, and Nancy are actually full siblings. Only the third generation reveals that Bella is highly inbred.

Kennel- and farm names are helpful in this respect, although you can't always depend on them to tell you the full genetic story.

Different ownership does not necessarily translate into linecrossing. If owner "A" begins his program with linebred animals closely related to those owned by "B," the two breeders more than likely do not have, genetically speaking, two different *lines*.

The first outbreeding can be a humbling and somewhat painful experi-ence for the novice breeder — especially one who enjoyed early success due to the lucky choice of excellent foundation stock or the smart advice of a knowledgeable mentor.

Flush with triumphs gathered in the show ring or at performance events, the novice is firmly convinced that his animals are far superior to anything else "out there" — but he will eventually realize that he must "re-group." He must use another breeder's stock. (As the years pass, he will mellow considerably and come to appreciate the diversity available in his breed, and the benefits one might gain from it.)

Where should you look? If you are truly enamored of what you have in your own kennel or barn, then search for a line or individuals closely resembling your animals in phenotype. By now, you should realize that the perfect animal does not exist — therefore, expand your hunt to include an animal blessed with superior quality in a trait your own line does not excel in.

Of course, all breeders have agendas with emphasis on certain criteria — and it's a good thing that they do. They create the genetic diversity so crucial for the survival and genetic gains of a breed or species. The day you decide your stock needs no improvement is the day to quit breeding altogether.

There is another breeding technique used to create new breeds, re-create or give a genetic "boost" to old breeds, and conduct genetic research. You'll learn about all this in the next chapter.

Chapter Twenty

CROSSBREEDING

CHAPTER TWENTY

Crossbreeding

"Crossbreeding: the mating of animals of different breeds, usually done to take advantage of hybrid vigor or to produce offspring with a potentially useful blend of traits from the parents."
D. Phillip Sponenberg, D.V.M., Ph.D., *A Conservation Breeding Handbook*, 1995

Why would anyone feel the need or desire to create new breeds? The sheep industry supplies us with excellent examples to support crossbreeding.

The Columbia sheep began as a cross between a Lincoln ram and a Rambouillet ewe. Breeders hoped for and achieved an increase in carcass weight and wool. The Corriedale, initially developed in Australia and New Zealand, resulted from Merino-Lincoln crosses. French breeders crossed several milk breeds to increase milk production.

These are just a few examples of crossbreeding in the sheep industry — not all crosses were successful. In *Raising Sheep the Modern Way* (1989), Paula Simmons reports on a breeding experiment at South Dakota State University. The objective was to create a sheep breed with no tails. Most sheep are born with long tails. They are docked to avoid accumulation of manure and to make breeding and shearing easier. The program was only moderately successful and was finally disbanded after fifty (yes, fifty!) years.

It is understood by most breeders that after individuals belonging to two different breeds are crossed, the offspring is bred back (*backcrossed*) to their sire or dam. There are exceptions. For example, the commercial livestock industry takes advantage of such crosses without the subsequent backcrossing. Their objective is not the creation of a new breed. They are primarily interested in the cross that produces the best product for their market. In other words, the result of the initial cross is an end in itself, and the cross is repeated over and over again.

Sometimes breeders have no choice. A good example is the breeding of mules, the sterile crosses of male donkeys (jacks) and female horses (mares). Their chromosome count is close enough for such a mating to produce offspring, but makes the breeding of subsequent generations almost impossible. There are, for all practical purposes, no "purebred" mules. Leah Patton, who is the office manager of the American Donkey and Mule Society, tells us in *American Livestock Magazine* (June/July 2000) that "... there is some scientific evidence that the rare female mule or hinny — with chances smaller than one-in-a-million — has actually foaled a living offspring."

A *hinny*, I learned, is the product of a male horse (stallion) bred to a female donkey (jennet or jenny). According to Patton, they are more rare than mules because stallions are usually reluctant to breed jennets, while jacks apparently service mares quite enthusiastically. Strange!

When species become extinct, they are forever lost to our ecosystem. Specific breeds within a species may also die out, and with their demise comes the complete loss of their unique genotype and phenotype, depriving our planet of yet another sliver of genetic richness and diversity. Members of conservation groups struggle mightily against such losses.

When confronted with the issue of extinct species or breeds, most people naturally assume one is talking about wild animals — tigers, cheetahs, and certain types of rhinos come to mind. Even astute environmentalists are often quite surprised to discover that domestic livestock and even pets have become extinct or are close to facing that fate.

Denise alerted me to a science article published in Reuters Press, *Biodiversity Shrinks as Farm Breeds Die Out* (September 22, 2001). It states, in part: "Breeds of farm animals are dying out and types of plants disappearing at an alarming rate, threatening long-term food security and depriving the world of their ability to resist disease and harsh climates." The article goes on to say that two breeds of farm animals disappear each week, and 1,350 breeds face extinction. Over the past 15 years, 300 out of 6,000 breeds of farm animals identified by the Rome-based U.N. Food and Agriculture Organization (FAO) have become extinct. FAO members warn that 30 percent of the world's farm animal breeds are at risk of disappearing, and their valuable traits, such as their ability to adapt to harsh conditions, disease, drought and poor quality feed, could be lost as well. Ricardo

Cardelino, the FAO's senior officer for animal genetic resources, states: "Once you lose a genetic resource, it's gone forever."

Readers interested in rare livestock such as Leicester Wool Sheep or Randall Blue Lineback Cattle as well as unusual horses, sheep, swine, poultry, and others will find the address for the *American Livestock Breeds Conservancy* (ALBC) in *A Conservation Breeding Handbook* (1995), written by D. Phillip Sponenberg, D.V.M, Ph.D. and Carolyn J. Christman, D.V.M, Ph.D.

The ALBC lists breeds with fewer than 200 annual North American registrations and a global population of fewer than 2,000 as *critical*. It lists breeds with fewer than 1,000 North American annual registrations and a global population of fewer than 5,000 as *rare*.

Breeders of Suri alpacas should find this book especially interesting and helpful. The North American Alpaca Registry — rather hastily closed — no longer accepts imports, which severely curtailed the small Suri gene pool.

I believe that Suris are not a separate breed from Huacayas, merely the carriers of a distinct fleece variety. I appreciate and applaud the efforts of the North American breeders who try to preserve their uniqueness. (Sponenberg and Christman make a definite distinction between *preserving* and *conserving* a species, breed, or variety. I can't quite decide which term should apply here.) The above-mentioned book will give alpaca breeders plenty of food for thought. Unfortunately, profit and sound breeding practices are sometimes diametrically opposed goals. Financial reality rears its ugly head even for breeders with the best intentions. In *Purely Suri* (Spring 2001), Dr. Sponenberg's article gives practical advice on how to sensibly approach a Suri breeding program.

Crossbreeding is used sparingly in a conservation program. The objective is to conserve an already existing breed, not to create a new one. There is one particular situation when crossbreeding becomes a necessity to save a breed from certain extinction. Denise is presently committed to such an effort. It is interesting and beneficial to the reader to read her report.

Ingrid Wood and Denise Como

The Tale of the Rampur Hounds of India

"In 1996, my husband and I recovered a small colony of Rampur Hounds, a nearly extinct sighthound of India, from Canada. The colony consisted of 7 adult dogs, and the youngest was in whelp. At home we finally sorted out what we had. Kulu and Taj were sisters. We had Kulu's offspring: full littermates Ganesh (male) and Bombay (female). Their younger sister was Cintamani. Taj's offspring: Shankar (male) and Shakti (female). Cintamani had been bred to Shankar. Kulu, Bombay, Ganesh, Cintamani, and Shankar stayed with us, while Taj and Shakti went to live with very good friends.

All the hounds are descendants of Gandhi* and Indira*, a brother and sister pair who were born in India and imported to Canada in 1984. The third littermate, Mahatma*, was killed in an accident after arriving in Canada. Gandhi* and Indira's* sire and dam were Baaz and Ch. Bathsheba, both Indian-born (New Delhi) and unrelated (assumed, but not verified).

Gandhi* is everyone's sire. Bred to his sister Indira*, they produced Delhi. Bred to Delhi, they produced Kulu and Taj. Bred to Kulu, they produc-ed Ganesh, Bombay and Cintamani. Bred to Kulu's sister Taj, they produced Shankar and Shakti. Shankar was then bred to Cintamani, combining the offspring of the sisters Kulu and Taj. However, Cintamani subsequently lost both puppies at birth. The other Rampuri born in Canada had been lost to accidents, management problems, and other bad kennel situations.

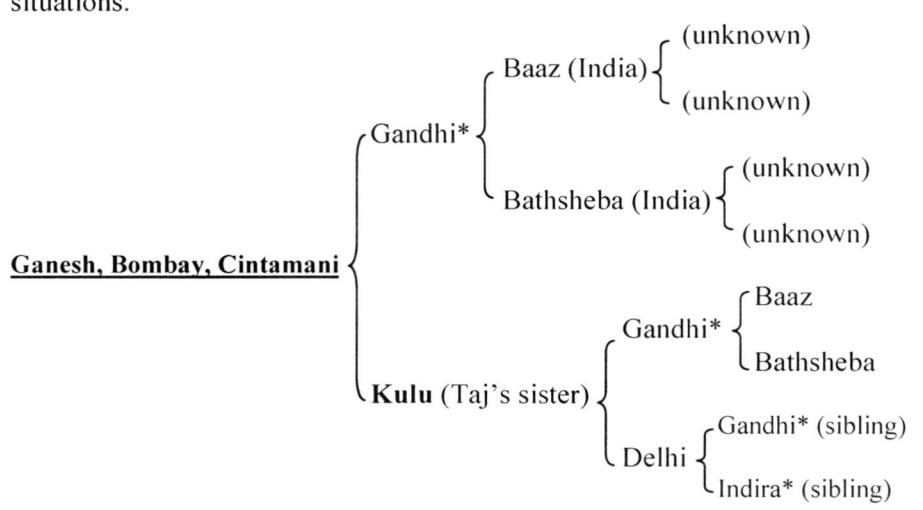

Ganesh, Bombay, Cintamani
- Gandhi*
  - Baaz (India)
    - (unknown)
    - (unknown)
  - Bathsheba (India)
    - (unknown)
    - (unknown)
- **Kulu** (Taj's sister)
  - Gandhi*
    - Baaz
    - Bathsheba
  - Delhi
    - Gandhi* (sibling)
    - Indira* (sibling)

A BREEDER'S GUIDE TO GENETICS
Relax, It's Not Rocket Science

(Note: an asterisk (*) after an animal's name in a pedigree, regardless of species, usually denotes an import into the U.S.)

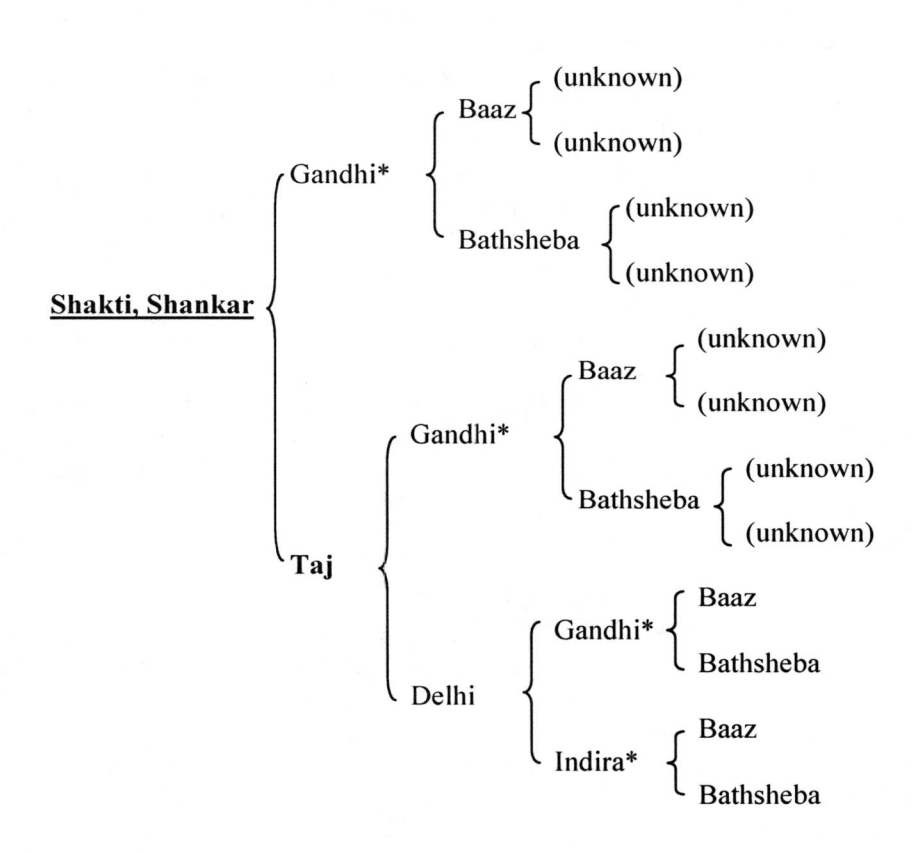

Shakti, Shankar
- Gandhi*
  - Baaz
    - (unknown)
    - (unknown)
  - Bathsheba
    - (unknown)
    - (unknown)
- Taj
  - Gandhi*
    - Baaz
      - (unknown)
      - (unknown)
    - Bathsheba
      - (unknown)
      - (unknown)
  - Delhi
    - Gandhi*
      - Baaz
      - Bathsheba
    - Indira*
      - Baaz
      - Bathsheba

I don't believe you can manage to find a smaller gene pool than that of our Rampur Hounds (it's barely even a puddle). During the course of the last three years, we lost Bombay (to renal failure), Ganesh (to multiple seizure activity), and Cintamani (to a severe temperament disorder — she also began suffering from vaginal prolapse during estrus). Shakti has undergone surgery for a mast cell tumor and has had chemotherapy — effectively taking her out of the breeding program. Taj was diagnosed with mammary and uterine tumors and was spayed. That leaves us with Kulu (who is over 10 years old and has several mammary "nodules") and Taj's son Shankar (who is nearly 9).

Rampur Hounds, never popular or numerous in their native country, are now nearly extinct. The influx of English Greyhounds during the British occupation of India further diluted the gene pool of the few dogs that were left. Attempts to import additional Rampuri have been unsuccessful. Attempts to produce offspring from our own dogs have, so far, been fruitless."

Are Denise, Richard, and their supporters just dreamers who have set themselves an impossible task? Is it possible to recreate a breed once it has nearly disappeared? For an answer, we have only to study the Irish Wolfhound. In ancient times, ownership of these swift, giant hounds was reserved for nobles and kings. The "cu" were used in battle and to hunt deer, boar, and wolves. In addition, they ably protected their master's property. As early as the 4th century, Irish Wolfhounds were sent as gifts to Rome. Between these exports and the demise of the wolves and giant stags in Ireland (to say nothing of the kings), the dogs adored by the Irish poets virtually ceased to exist. In *Windhunde* (1979), Ingeborg and Eckhard Schritt, who admiringly call the Wolfhounds *"Aristokraten,"* credit George Augustus Graham (1833-1909) with saving the breed.

How was this accomplished? The remaining Wolfhounds were crossed with Scottish Deerhounds when necessary. Some breeders also tried Wolfhound to Great Dane crosses, but these did not produce good type.

In Denise's research, she found that Captain Graham wrote (in the 1908 *Kennel Encyclopedia*, Vol. II, edited by Sidney Turner): "Richardson in 1841 got together as many Irish Wolfhounds as he could and continued the breed, which Sir John Power, of Kilfane, kept up; and he, Mr. Baker of Ballytobin, and Mr. Mahoney of Dromore were the last Irishmen who really tried to keep up this magnificent breed. In the year 1862, the writer [Captain Graham] took up the breed and since then, his life has been devoted to it. Fortunately, Sir John Power was a friend of his [Captain Graham], so he started with the purest possible blood of the Kilfane and Ballytobin strains." Captain Graham then added: "The present breed of Irish Wolfhounds has been built up by bitches obtained from these two kennels, crossed with the Scotch Deerhound, a very similar but much slighter dog ... yet the writer confidently believes there are strains now [1908] existing which may be traced back, more or less clearly, to the original breed."

A BREEDER'S GUIDE TO GENETICS
Relax, It's Not Rocket Science

This swift Scottish Deerhound spends her weekends lure coursing. Her prey? White plastic kitchen bags! Lure coursing is enjoyed by sighthounds and their owners without using live game, betting, or winning cash prizes. Hounds retired from the field remain with their families as beloved companions
(photo by Barbara Ewing)

Does Denise have a plan for the Rampur Hounds? She resumes...

"We continue our worldwide search for other Rampur Hounds. However, we can speculate that the last generation of our hounds, the ones who have had the serious health issues, are possibly one generation too tightly inbred and/or linebred to remain viable as healthy breeding specimens. Our option is to breed the remaining Rampur Hounds to hounds of similar type and who still have very strong hunting instincts — for instance, to smooth Salukis of desert (Bedouin) lineage, to Sloughis (African gazehounds), even possibly to Chart Polskis (Polish gazehounds). We would need the best possible phenotypes to keep the outward appearance of the drop-eared, heavily muscled, powerful Rampur Hound from being lost to a lighter, sleeker, rose-eared Greyhound type.

After breeding various combinations of the above dogs to our Rampur Hounds, we would then breed those offspring together, bringing the concentration of Rampuri back into the program. As an example:

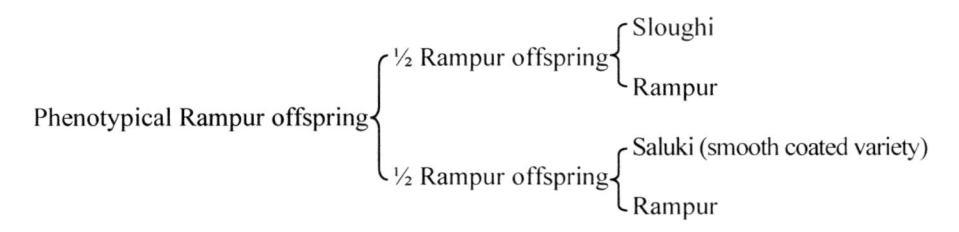

Phenotypical Rampur offspring
- ½ Rampur offspring
  - Sloughi
  - Rampur
- ½ Rampur offspring
  - Saluki (smooth coated variety)
  - Rampur

The influx of Sloughi and/or Saluki genes would give the Rampur offspring much-needed hybrid vigor, but still stay close to the phenotype and strong hunting instinct we desire. However, such a plan would require many breedings, and a number of dedicated individuals willing to continue with such a program."

Denise realizes that the genetic interpretation of the above pedigree is very simplistic in its presentation. Why?

We already discussed the concepts of the *Law of Segregation* and *Mendel's Law of Independent Assortment*. These ominous sounding terms describe the "splitting up" and the random distribution of parental genes (alleles). When genes cross over during meiosis, things are re-shuffled even further.

As we discussed much earlier in this guide, each process of cell meiosis actually results in the formation of four (haploid) cells, although only one of the four female cells is viable. Germ cells produced by the same animal can show quite a bit of genetic variation.

When we sort the phenotypes of Denise's pedigree, we can form combinations of Rampur-Rampur, Rampur-Saluki, Rampur-Sloughi, and Saluki-Sloughi. "Oh, come on!" you exclaim. "These combinations look pretty simplistic themselves!"

You are correct — we need to take this a little further. The authors of *The Nature of Life* present an easily understood example by inviting the reader to visit a restaurant. Suppose a restaurant offers six courses per meal. Each course consists of two choices (homologous chromosomes). You can arrange 2^6 or 64 different meals (genomes) from such a selection. Now replace the 6 by the number of chromosomes in sperm or egg cells. Get out your calculator!

A BREEDER'S GUIDE TO GENETICS
Relax, It's Not Rocket Science

The often-read statement that an animal receives 25 percent of its genome from each grandparent should not be taken literally. Let me explain.

		AA
	Aa*	
		a*A
a*a*		
		a*a
	a*a	
		aA

In our example, the letters may represent one of the many polygenes controlling size, for instance, in an alpaca. Only two of the four grandparents (marked *) make genetic contributions at this locus. You can plainly see that the paternal grandsire and the maternal dam have no part in this particular <u>tiny segment</u> of the baby's genome. Their alleles (**A** and *a*) residing at this <u>specific locus</u> are permanently lost to future generations. You must visualize this concept for the entire genome of an animal. Which two genes get to "move ahead" to the next generation is, of course, left to chance.

In *The Animals in My Life* (1996), Grant Kendall describes how his pet mallard duck Daisy mated with a large white Pekin. Several weeks later, twenty-four ducklings hatched. According to Kendall, "They were a strange lot. Some looked like little mallards and some were totally yellow, like baby Pekins. But most had varying shades and amounts of black and yellow, as they were a blend of their two parents." The mallard duck, by the way, is a distant ancestor of the Pekin.

139

The truth is that you might not have inherited a single allele from the distant royal ancestor or Mayflower passenger you were so proud to discover in your "lineage." The great-grandson of the sheep that won the blue ribbon at the Farm Fair might only be marginally related to his famous great-grandsire. Now you know why astute breeders used linebreeding programs long before the concept of the double helix was conceived and proved correct.

It all makes sense to you now, doesn't it? Does your son look like the image of your father? Does your daughter have your mother's body type but your husband's facial features? They are as much a product of *allelic segregation* and *independent assortment* as the squealing puppies in their whelping box or the newborn foal nuzzling its mother in their stall. It is rare that siblings carry an almost identical genome.

So, you see the statement that "A child inherits 50 percent of his/her genes from each parent" is perfectly true, but no two children inherit the same 50 percent. Mathematical concepts are always easier to understand when we shrink the numbers involved. Picture each parent holding 100 marbles, and the child chooses 50 from dad and 50 from mom — then imagine the potential combinations. A brother-sister mating may genetically be a true outcross! Although such an occurrence is statistically highly unlikely, it is possible.

To return to our Rampur Hound Project: the distinct possibility exists, as Denise showed in her pedigree, that at least one or two puppies would show a phenotype closely resembling that of the two Rampur grandparents. Denise again cautions that "such a plan would require many breedings."

Crossbreeding, outbreeding, and linecrossing can create what geneticists call *hybrid vigor*. Novice breeders unfortunately often harbor totally unrealistic expectations from this concept. We need to explore it for better understanding.

Chapter Twenty-one

HYBRID VIGOR

CHAPTER TWENTY-ONE

Hybrid Vigor

"Heterosis occurs when the progeny exceed the mid-parent average by a statistically significant amount."
Malcolm B. Willis, 1989

The concept of hybrid vigor, called *heterosis* by geneticists, can be a tricky one to grasp.

Inbreeding and/or linebreeding can result in *inbreeding depression*. Overall health, disease resistance, and fertility are the areas most affected by inbreeding depression in many species, including human beings. Simply put, inbred or linebred individuals often lack *vigor* (energy, vitality, strength). Inbreeding or line breeding can also speed the process of bringing recessively carried genetic defects to the surface.

I bet you're waiting for me to cite my customary exception — I am happy to oblige. Several researchers created as many as twenty or more generations of inbred rats (in the case of Dr. Helen Dean King more than one hundred), using brother to sister matings. The rats grew larger and more vigorous with each successive generation. But, and this is important to remember, the researchers only selected the healthiest specimens from each litter to continue their breeding program.

Some authors drag out the traditional history of the old Egyptian royalty with their brother-sister marriages resulting in healthy offspring over many successive generations. Let's get a grip on reality here! Just because the royal siblings married each other doesn't mean they produced offspring together. I doubt the average commoner was informed of the true identity of each "royal" baby's biological father. The ignorant masses were kept in the dark, so to speak — and the power was kept in the family. The sperm "donors" may have been "done away with" to preserve the secret. I am sure similar shenanigans were enacted among the "royals" of Europe, although close genetic alliances undoubtedly resulted in offspring. One only needs to contemplate the regular appearance of the famous Habsburg jaw and various genetic disorders among the "blue bloods."

Dr. Willis gives the following definition for heterosis: "Strictly the situation when the progeny performance exceeds the mid-parent performance in a cross between two distinct lines." He uses the height measurements of adult males as a simple example. Let's assume you cross two breeds in which adult males average 64 and 62 cm in height at the withers. The midpoint between 64 and 60 cm is 62 cm. A measurement of 63 cm has therefore "exceeded the midpoint and obtained what is called hybrid vigour or heterosis" (Willis). Although Dr. Willis chose the inheritance of height as an example, he points out that "traits concerned with growth, weight, carcass composition and the like are not generally associated with much heterosis."

Under Dr. Willis's definition of heterosis, the offspring's average does not have to exceed that of the superior parents — as long as it surpasses the midpoint measurement. Regardless of the species you are interested in, I strongly suggest that you study Willis's explanations in *Genetics of the Dog* for greater clarification.

Blood and Studdert defines heterosis as "greater vigor than is shown by either parent."

It's clear that the concept of heterosis does not apply equally to all traits. Before you read on, go back for a quick refresher on polygenic inheritance in Chapter Thirteen. Low heritability traits such as general health, disease- and parasite resistance, as well as fertility, often show great improvement in a hybrid. Additive traits of high heritability, such as coat- or fleece quantity and quality, are not positively influenced to any large extent under such a program.

Crossbreeding two individuals with <u>poor</u> coat (fleece) quality will not produce a great coat or fleece due to heterosis. It is true that luster or even density might marginally improve due to an overall improvement in health, but do not hope for substantial gain under such circumstances. You can't press blood out of a turnip, as the saying goes. In *Genetics of the Dog*, Dr. Willis clarifies the entire concept very well when he tells us not to expect much heterosis "in those traits known as luxuriant traits, that is to say, not directly crucial to the future of the species."

Dr. Willis wasn't talking about human selection pressure, but about traits that help animals to survive and thrive in the wild. Examine the animals you plan to breed with that thought in mind.

In other words, outcrossing or crossbreeding do not perform miracles in, for example, the improvement of density or luster in sheep fleeces. The breeder of sheep with average fleece density must choose mates with <u>superior</u> density to improve his stock. In contrast, a ewe with overall marginal health will often give birth to robust offspring if bred to a ram genetically unrelated to her, even if this ram is not the most robust specimen himself.

What about the elimination of defects, often a major reason breeders resort to outbreeding and whose absence certainly promotes health and vigor? You must understand that the elimination of defects will only be achieved if a difference in gene frequency exists between the parents. The phrase "difference in gene frequency" is easily explained by using the example of coat color (although coat color is only indirectly connected in any way to the concept of heterosis, as discussed in this context).

If I breed two **AA $E^m E^m$ BB DD SS** (black) Whippets, there is no difference in gene frequency for color. (Don't be thrown off track by all those letters. You will fully understand why I used them by the end of the book.) The dam could be American bred and the sire an English import with the pedigree not showing a common ancestor in thirty generations — but only homozygously black puppies will be produced. Likewise, if your stock suffers from the frequent appearance of some horrendous defect like *pyruvate kinase deficiency* or *progressive retinal atrophy*, outbreeding or linecrossing will not help to eradicate it if the animals you breed your stock to carry those particular defects themselves.

Writing for *The Aristocrat* (Borzoi Club of America, Fall 2000), Dr. Jerold S. Bell made this point: "Repeated outbreeding to attempt to dilute detrimental recessive genes is not a desirable method of control. Recessive genes cannot be diluted; they are either present or not."

The very act of outbreeding or crossbreeding does not negate hereditary principles — be they simple or complicated. Genetically speaking, <u>a true outcross increases the number of heterozygous allelic combinations in the offspring</u>. If it doesn't accomplish that, the "outcross" exists only on paper.

Oddly enough, the autobiography of former president Ronald Reagan offers a perfect example. When President Reagan visited Ballyporeen in Ireland, a young man in his early twenties was introduced to him as a distant relative. "It was amazing how much he and I resembled each other — his eyes and hair, his whole facial structure all resembled mine. Since it had been over a century since my great-grandfather had left Ballyporeen for America, it was an eerie experience" (*An American Life*, 1990).

So, the stud you import from Germany to introduce new genes into your Boxer line might be genetically closer to your stock than an animal located in the next county. Sometimes, there is just no way of knowing this (although in-depth research helps eliminate some surprises).

Instead of ridding your line of defects, you can just as easily introduce new ones through such a program. Australian German Shepherd breeders brought pituitary dwarfism into their kennels when they outcrossed to English and German lines, after a forty-four year import ban was lifted. Instead of hybrid vigor, such outcrossing resulted in hybrid misery. The crossings of Irish Setters with Standard Poodles at a colony maintained by the University of North Carolina resulted in cases of subluxation (dislocation) of the carpus (wrist-bones in the foreleg).

Outcrossing is not always the sure path to health and happiness that novice breeders think it is. It does not improve <u>all aspects</u> of a breeding program, and under certain circumstances might not improve any. Breeders also need to understand that when hybrids are bred back into the original line, any heterosis accrued will disappear rapidly.

<u>While breeders should not expect to reach unrealistic goals from outbreeding, they are wise to remember that hybrid vigor (heterosis) only occurs with increased heterozygosity</u>. Any time your breeding goals call for "uniformity," you are striving to increase homozygosity in your stock — while decreasing hybrid vigor. Breeders need to understand this before advocating breed standards that demand uniform conformation. Artificial insemination, used on a large scale, also promotes genetic uniformity — read homozygosity — in a population. We have already discussed the danger of narrowing gene pools (resulting in genetic bottlenecks) to species not bred for slaughter.

Let's look into defects and how they impact the concept of heterosis.

Chapter Twenty-two

GENETIC ANOMALIES

CHAPTER TWENTY-TWO

Genetic Anomalies

"Anomaly: a marked deviation from normal."
Blood & Studdert, *Baillière's Comprehensive Veterinary Dictionary*, 1988

A few months ago, a colleague handed me a photograph accompanied by a newspaper article. After reading the article and studying the photo, I sat in total shock, tears brimming in my eyes. The well-written piece described the heartbreak felt by my co-worker's friends upon learning that their only child was dying of Tay-Sachs disease, a genetic illness primarily suffered by members of the Jewish population.

My interest in genetics familiarized me with the disease years ago. What was so special about this case? The parents were not Jewish. They were of German and Irish descent, two ethnic groups normally not associated with this defect. The baby, still healthy looking and smiling in an old photo — well, this baby resembled my own son closely enough to be his twin. The same white-blonde hair that covered my Ben's head as an infant and toddler, the same shape to the nose, the ears, shape of the head, that same sturdy build — he could have been my child. My husband and I are, respectively, of Irish-English and German ancestry.

Years ago, one of my former students, a brave and stoic little guy, died of Sickle Cell Anemia, a disease affecting Africans and members of the African-American community. Cystic Fibrosis strikes one in every sixteen-hundred Caucasian babies. Many more are carriers. The estimate that geneticists Ian Wilmut and Keith Campbell give is one in twenty people.

What do these and many of the other estimated four thousand genetic defects afflicting humans have in common? They are carried as recessives.

You learned in the chapter on dominant *vs.* recessive traits what the odds are of a recessive allele making its appearance. You also understand now that the offspring of two carriers may express the actual defect. The good news is that scientists have identified the genes responsible for defects such as Tay-Sachs, Sickle Cell Anemia, and Cystic Fibrosis. A simple blood test administered to both potential parents can identify them as either carriers or non-carriers of Tay-Sachs or the Sickle Cell trait. (In the case of Cystic Fibrosis, it's a little more complicated, but identification is nevertheless possible in most cases.) If two such carriers produce a child, it may exhibit the illness, it may be born as a carrier itself, or it may be lucky enough to inherit the two healthy alleles.

Let's review. **D** = non-defective allele, **d** = defective allele.

	D	d
D	**DD**	**Dd**
d	**Dd**	**dd**

Remember, these are probabilities! If the baby is born **DD** (which letter we use to explain the concept is not important), the fortunate joining of the two dominant alleles means this child's *genome* is and remains free of all traces of this particular defect. The probability of two carriers producing a child with the defect is 25 percent.

Upon learning that one or both people are carriers, prospective parents can weigh the risks before deciding to have children.

Oddly enough, scientists must also worry about possible negative repercussions of eliminating various anomalies. How so?

Nature, working in the strange ways that she does, actually tolerates the propagation of certain defects to prevent other problems. Those suffering from Sickle Cell Anemia, as well as carriers of the trait, are protected from the ravages of malaria. Caucasian explorers of the African continent, who carried no such "protection" in their blood cells, perished in large numbers in what was for them an unforgiving climate.

Researchers now speculate that carriers of the Tay-Sachs gene might have been protected from the effects of tuberculosis, a disease that swept through all of Europe for centuries. It was especially prevalent in the crowded cities, including those sections of the densely populated Jewish ghettos.

A defect can be of genetic origin, can be caused by environmental influences such as poor nutrition or the ingestion of poisonous plants, or can possibly be the result of a combination of both. Spontaneous mutations or those triggered by chemical or radiation exposure may also contribute to abnormal development. In *Medicine and Surgery of South American Camelids* (1998), Murray E. Fowler, D.V.M., adds trauma and hyperthermia to this list.

Animals suffer from so many defects that their study often leads a breeder to despair. How is one to purge one's breeding stock completely of all the evil recessive alleles lurking in hiding places? Many breeders — of all species — recognize the importance of medical research and give generously of their time and money to support it.

Genetic research is time consuming and therefore expensive. Breeders need to choose their battles carefully. Common sense dictates that funding should first be applied to search for those DNA markers leading to "lethal" genes (meaning any gene leading to certain death either in utero or shortly after birth). Next, funding should be provided for identification of genes causing severe physical impairment, making it impossible for an animal to function without assistance. Less threatening defects such as cryptorchidism could then be investigated. This condition requires castration due to the possibility of retained testicles becoming cancerous, but does not otherwise impair the animal's quality of life. (Since most breeders will not use a cryptorchid at stud, some might argue with my "quality of life" evaluation.) Finally, defects requiring minimal or no veterinary or owner intervention could be studied.

Llamas and alpacas are generally healthy and easy to care for, but they do suffer from their share of problems, just like any other species. One of the nastiest defects affecting them is *choanal atresia*, a lack of passageway between the nasal cavity and the throat. Babies are unable to suckle and breathe at the same time. Corrective surgery is seldom feasible. The cria must be euthanized (humanely killed), or it will die of starvation.

Other severe problems plaguing alpaca breeders are *atresia ani* (no anal opening), cleft palate, lethal heart murmurs, and impaired brain development, leaving the cria unable to walk. These defects are rare at this time, but are devastating when they occur.

Following these is a host of other non-lethal defects such as undescended testicles, malocclusion (misalignment of teeth or jaw), small umbilical hernias, closed tear ducts, and agalactia (poor milk production). They nevertheless require surgical intervention or extensive maintenance.

While it is true that many blue-eyed white alpacas are deaf, they do not require medication or special consideration. They can function well as herd members and, from personal observation, do not pass this defect on to their <u>colored</u> offspring (more about this in the chapter on color genetics).

Most camelid defects are only suspected to be of genetic origin at present. Breeders who believe, however, that only environment issues play a role are simply deceiving themselves. A species without genetic defects does not exist.

You as the breeder must decide which defects you (and your animals) can live with. Prioritizing selection criteria can be a gut-wrenching experience for those who put much love, care, and thought into their breeding programs.

About 350 hereditary diseases have been identified in the dog world so far. In the 1999 survey conducted by the AKC Canine Health Foundation, breeders listed epilepsy, hip dysplasia, cancer, bloat, and hypothyroidism as their top concerns.

I've mentioned before that some defects, such as cardiac problems or hip dysplasia, are inherited through polygenes (a group of genes). If enough of these genes come together (genetic loading) in one animal, the defect will be expressed. We label such genes *threshold characters*.

Many pet and livestock breeders realized long ago that genetic anomalies are part and parcel of breeding. They collectively took their heads out of the sand and aggressively pursued the search for DNA markers, as well as for specific genes coding for defects — with great success, I might add! One needs only to read the breed columns in the *AKC Gazette* magazine on a regular basis to keep abreast of research and triumphs over genetic problems. Much work still needs to be done, of course, as the chromosome locations of many defective genes are still mysteries. A breeder of any species who says that his or her line is free of all anomalies would cause an eruption of derisive comments among experienced and ethical breeders. Such a boast should also send a potential buyer running to the nearest exit!

If you eliminate all breeding stock compromised in any way by genetic defects, including all carriers and their siblings (as some authors suggest), you simply will not have any animals left to breed. Hence, my previous suggestion to choose your battles wisely. Breeders of all species must cooperate to set <u>sensible priorities</u> as far as the elimination of defects is concerned. It can be done. What am I saying? It has been done!

Chapter Twenty-three

DNA TESTING

CHAPTER TWENTY-THREE

DNA Testing

"Deoxyribonucleic acid: a nucleic acid of complex molecular structure occurring in cell nuclei as the basic structure of the genes. DNA is present in all body cells of every species, including unicellular organisms and DNA viruses."
Blood & Studdert, *Baillière's Comprehensive Veterinary Dictionary*, 1988

In *The Roots of Life* (1978), author Mahlon B. Hoagland likens each individual to a "runner in the relay race where DNA is the baton." Beautiful! Picture a succession of your animal's ancestors over a span of hundreds or thousands of years — each generation passing the double helix to the next one. Of course, if you paid attention to the information in previous chapters, you know that genetic content changes somewhat with each passing of that baton.

What absolutely amazes me and every other non-scientist is the fact that the mindboggling diversity of the earth's species is created by reshuffling just four related molecules called *nucleotides*. The peasant farmers of South American call llamas and alpacas their "speechless brothers." Those pastoralists intuitively seem to grasp a genetic concept the average "educated" person in this country is not aware of.

Let's review: four links (nucleotides), the scientific names for their differing bases abbreviated to A, T, C, and G (memory bridge: All Ticks Carry Genes — sorry, I couldn't think of a more catchy phrase), are the only "letters" used in the genetic "book" called DNA, whether they describe a llama, a pig, a hamster, or a human being. What accounts for the uniqueness of each individual? The difference is how the bases of these four nucleotides are sequentially arranged.

More review. One of the key "ingredients" in a human or animal is protein. At one time scientists were convinced that genes were composed of protein. We now know that protein is created by instruction from DNA. All genetic information is stored in DNA; enzymes copy each gene and turn it into a molecule called *messenger RNA* (*ribonucleic acid*). The RNA molecules carry the genetic message from the cell nucleus to the cytoplasm, the cell's protein factory.

Lisa Foreman, manager of DNA programs at the National Institute of Justice, calls DNA "God's gift to forensics," as it helps to convict the guilty and release the innocent from prison (as quoted in *The Philadelphia Inquirer*).

DNA tests make it possible for the administrators of animal registries to identify breeders who are guilty of carelessness — or deliberate fraud — in their breeding programs. The *AKC Gazette* cancelled forty-four litters in a single month (August 2000), as a result of their DNA Audit Program.

At the Quarterly Meeting of AKC delegates in September 2000, President Alfred L. Cheauré reported a steady decline of litter registrations from commercial breeders once the Frequently Used Sire DNA Program was put in place. (Yippeeee!) Readers can draw their own conclusions.

DNA testing serves another important purpose for breeders of all species.

In the past, flushing out a defect and establishing the identity of carriers involved a very timeconsuming and costly process. Suspected animals had to be *test bred* to each other, their offspring raised, and sometimes, upon confirmation of a breeder's suspicions, euthanized. This took a terrible toll on a breeder's financial and emotional resources. Prominent studs were often used hundreds of times before being implicated in genetic disasters, occasionally devastating a breed's entire gene pool. Others were falsely accused, based on multiple appearances in certain suspect pedigrees. Breeding programs were ruined, friendships and reputations destroyed by rumors and poorly understood genetic concepts.

We are so lucky today! Scientific technology is increasingly available to breeders, taking more and more of the guesswork out of the search for defective genes. Breeders must cooperate with each other — and with scientists — to fund the continuation and expansion of such research. Several diagnostic tests are currently being offered to dog and livestock breeders. In the June, 1999 issue of the *AKC Gazette*, the columnist for the Shih Tzu breed proudly announced the discovery of the genetic marker linked to *Juvenile Renal Dysplasia.* The gene for *Von Willebrand's Disease* is detectible in Pembroke Welsh Corgis, Doberman Pinschers, Shetland Sheepdogs, and others. There are a number of additional tests for other dog breeds as well as for other species. Interested breeders might want to visit the VetGen web site listed in the reference section for further study.

We must differentiate between linkage-based tests and gene-specific DNA diagnostic tests. Only the gene-specific tests are 100 percent accurate.

In linkage-based tests, scientists identify the genetic marker that is close to a disease gene on a chromosome. It is important to understand that the genetic marker itself is <u>not</u> the disease gene. During chromosome crossovers, the marker and the disease gene can become separated and result in false negatives during linkage tests. Individuals may also carry the marker gene, but not the mutated gene causing the defect. In that case, the linkage test leads to a false positive. It becomes clear that only gene-specific DNA tests offer total assurance that an animal is not a carrier. In those tests, the defective gene itself is identified. Genes inherited in a simple Mendelian pattern are, of course, much easier to locate through DNA testing than those leading to a disease involving multiple genes — such as the previously mentioned heart problems or hip dysplasia.

Genetic diversity offers protection from the frequent joining of carriers whose DNA codes for a specific defect. Do not confuse the concepts of <u>population size</u> and <u>genetic diversity</u>. As strange as it sounds, there are species or breeds numbering in the thousands that carry a very limited assortment of alleles at each locus. The actual result is a small gene pool (Denise reminded me that this happens in dog breeds more often than people think). We need to dig up the old mitten example. Instead of having its members carry combinations of black, white, yellow, red, and green mittens (two per animal) at one genetic address, a population might quite possibly have its choices reduced to only red and green or, worse yet, only red.

A good example of a <u>large population</u> linked to a <u>small gene pool</u> with a high frequency of problems is the Doberman Pinscher. According to George J. Brewer, MD, one of the originators of VetGen, only 20 percent of Dobermans are clear of Von Willebrand's Disease (bleeding disorders); a whopping fifty percent are carriers.

Genetic counselors do not advocate the removal of all carriers from a gene pool. If their caution does not seem to make sense, think about it! If your animals belong to a species or breed with an already tiny gene pool, removing large numbers of them from the general breeding population will concentrate the available genetic material even more, and lead to what geneticists and breeders refer to as a *genetic bottleneck*. In your haste to rid yourself of one problem, you might very well create others.

Why use DNA tests under such circumstances? They identify carriers of the defect and those animals that are clear. The prudent plan would be to mate carriers to clear animals. If you remember our Punnett Square, the <u>probability</u> of getting clear offspring out of such breedings is fifty percent. Any progeny used for breeding must be tested for carrier status. Carriers are only used if their other contributions to the breed or species are outstanding and their elimination from the gene pool would be harmful to the population.

The goal, of course, is to <u>replace</u> carrier parents with their clear offspring in each subsequent breeding. <u>Selective</u> removal of carriers is the key to success. Using such a program, a population can be slowly purged of all carriers and, eventually, the disease itself. In the meantime, genetic diversity has been preserved.

The human genome, with its thousands of genes (I've read so many different estimates that I'm afraid to quote any concrete number), has been "mapped" through the collaboration of scientists all over the world. As I write this, work on sequencing the human DNA "text" is underway. We are on the verge of a genetic revolution with all its exhilarating — and frightening — repercussions.

While scientists around the world have been mapping and sequencing the human genome, those who devote their professional lives to helping animals and their owners are not idle. Researchers have been and still are actively working on genetic mapping and sequencing projects for dogs as well as agricultural animals.

As of 2002, 18,000 genetic sequences were identified for the dog, 115,000 for the pig, and 240,000 for the cow. The Alpaca Research Foundation (www.alpacaresearchfoundation.org) and the Morris Animal Foundation are providing the funding to map the alpaca genome. No animals will suffer to bring this ambitious project to fruition. Warren E. Johnson, Ph.D., principal investigator at the Laboratory of Genomic

Diversity, Maryland, used a licensed veterinarian to take a skin biopsy from an anesthetized male alpaca.

Dr. Johnson's research will open the door to locating and eventually sequencing the genes responsible for many of the congenital defects known in camelids. Identification of genes coding for economically important traits such as wool quality will make those traits more accessible to study.

Occasionally, genetic traits appear spontaneously as the result of a mutation. We'll explain this in the next chapter.

Chapter Twenty-four

MUTATIONS

CHAPTER TWENTY-FOUR

Mutations

"Mutation: 1. A structural alteration in DNA present in a mutant that gives rise to the mutant phenotype, and 2. An animal exhibiting such a change; a sport."
"Mutant: 1. A normal organism that is different in one or more characteristics from an arbitrarily defined, previously existing 'wild type' organism, and 2. produced by mutation."
Blood & Studdert, *Baillière's Comprehensive Veterinary Dictionary*, 1988

Many scientists believe that evolutionary changes are the result of *mutations*. Humans still share a percentage of identical gene frequences with animals as phenotypically remote from us as a mouse.

Mentioning the little rodents reminds me of an interesting tidbit I read in *Colour Inheritance in Fancy Mice* (no date given, printed in England by Watmoughs Limited — the copy I found in a country store that sold used books and pigmy goats belonged to a Howard Stevens, who entered the date of 1939 below his name). The author, W. MacKintosh Kerr, tells his readers about a lethal color mutation that "... although not known to the Fancy before 1900, is really a very ancient one, for it has been recorded as early as the year 1100 B.C. in the first Chinese Lexicon."

A few years ago, an alpaca show judge mused on the Internet about the origin of the astonishing variety of alpaca fleece colors and patterns. They evolved from the uniformly cinnamon-colored fleece of the vicuña, the ancestor of the alpaca (white fiber is only found on small, very specific areas of the body). How did this happen? Researchers proved that domestication allows mutated forms of the original color genes to surface. In one study, for example, wild foxes produced offspring with black and white patterned fur after successive generations were selected for tame behavior.

Mutations resulting in drastic change of appearance (think of the achondroplastic legs of the dachshund or the distinct coat of the Rex cat) do not occur nearly as often as people think. Breeders sometimes explain poorly understood modes of inheritance as mutations (also called "sports). Mutations do appear spontaneously at times. Their very appearance helped genetic scientists in their research on chromosome mapping. Only one problem arose: the mutations did not occur often enough in the fruit fly *Drosphila* to suit the purposes of the geneticists.

Luckily, the scientist Hermann J. Muller recognized that mutation rates increased rapidly when fruit flies were exposed to massive doses of x-rays. He discovered that defects created under these circumstances were passed on to successive generations in a recessive Mendelian mode.

During my studies, I found two examples of spontaneous mutations mentioned with a fair degree of regularity. They both involved color inheritance — one the mutation of a recessive allele coding for white to dominant white (**W**) in horses, the other a mutation in certain breeds of dogs from the **e** allele, which allows no expression of black pigment, to the **E** allele, which does. The fact that both authors (Sponenberg and Little) distinctly mentioned these specific cases indicates, however, that they are rare occurrences in the overall scheme of suddenly altered phenotypes. Dr. Willis claims that spontaneous mutations are so rare (one in ten thousand) that they are "to be safely ignored in most cases." Mutations obviously occurred often enough to account for the amazing diversity of all living organisms on earth.

In *The Secret of Life*, Levine and Suzuki tell how, through a mutation that occurred millions of years ago, ruminants slowly developed the ability to survive on a diet severely deficient for other mammals. Ruminants carry a subtly altered form of *lysozyme* in one of their stomachs. This enzyme is normally only carried in the blood, saliva, and tears to eliminate harmful bacteria. It makes it possible for ruminants to efficiently digest grasses and leaves. Alpacas and llamas, although not ruminants in the strictest sense of the word (their stomachs only consist of three compartments rather than the four of other cud-chewing mammals), reign supreme in the efficiency department. Their ability to survive and procreate on what is considered sparse grazing, even for ruminants, is truly amazing. Notice my choice of the word "survive" — not thrive. We'll address that issue later.

How does such a mutation take place?

During DNA replication, mistakes can occur. Bases are left out, transposed, or replicated more than once. Such copying mistakes can result in defects. Geneticists tell us that occasionally, during the formation of an organism, extra copies of genes are created and then kept in storage, so to speak, like the many mysterious tools and gadgets in my husband's overflowing garage. They <u>seem</u> to serve no practical purpose (both the genes and the gadgets). Geneticists appropriately call these genes *"junk DNA"* (and in the case of the gadgets, just *junk*). Sometimes, as in the case of ruminants, these extra copies take on new and important functions.

Most organisms, including humans, carry ancient unused DNA in their genome — evolutionary baggage. These genes were possibly once useful and necessary for survival, but are no longer needed. Still present in DNA, they do not code for protein. We can compare this to machines that have become obsolete over time. Millions of typewriters sit in closets and attics — very few people use them these days. Of course, if Mother Nature was like the hoarder I married, we'd still carry the entire DNA sequence of *Australopithecus Afarensis* (one of our distant ancestors) in our genome.

Increasingly, scientists believe that junk DNA <u>does</u> serve a functional purpose, and they are exploring that facet of genetics more aggressively.

Mutations can be the result of radiation, as in our fly example. The source of such radiation may be man-made or occur naturally in the environment. Chemical exposure is a well-known causative agent.

Recessive traits that appear after generations of "hiding out" (*atavism*) are not to be considered mutations.

A fellow Whippet breeder's bitch gave birth to a little adorable female with a stubby tail, a most unusual occurrence in the breed. All the siblings sported the customary long Whippet tails. The breeder speculated whether the stubby tail was a mutation, or if the bitch could have possibly bitten the end of the tail off in her frenzy to open the puppy's placental sac to consume the placenta (as bitches will do). Who knows? The puppy may just have not developed completely while *in utero*. In this instance, we could describe the stubby tail as a *congenital defect* — meaning it was present at birth — rather than a *genetic defect*.

The common belief is that animals develop adaptations (mutations) in response to changing environments. Apparently, they don't. Geneticists tell us that, as environments change, only those animals that develop mutations that <u>happen</u> to favorably deal with these changes are able to survive. Lay people think this is splitting hairs, yet it presents a very important genetic concept.

Let's emphasize this again: specific mutations <u>do not</u> develop as a reaction to a changing environment. Once they appear, however, certain mutations make survival easier for individual organisms in specific environments. Sickle Cell Anemia, for example, did not develop in response to the threat of malaria. The first African whose gene mutated to produce the sickling trait was, coincidentally, protected against malaria, lived to adulthood, and was able to pass on his genes — along with the sickling trait. His offspring carrying the trait were thus favored for survival, and in turn were able to pass it on to their children.

A favorable mutation in one environment might lead to extinction of a species in another or changing environment. An example was presented by one author describing an insect's color mutation from white to grey. When pollution dirtied their environment, the grey mutants were not as easily spotted by birds looking for a meal. Eventually only the grey insects survived — protected by a mutation (*industrial melanism*) they had not actively brought about. Mutations carry a luck factor. They are not the result of good or poor "planning" on the part of an organism.

Of course, such views and research on evolutionary changes are not without their opponents and controversy. A virtual maelstrom of emotions swirled around the poor guy who presented the moth research. Judith Hooper's book *Of Moths and Men* tells a tale of "human ambitions, rivalries, weaknesses and scandals" (Michael J. Behe, *Philadelphia Inquirer*, 2002). Although one is tempted to joke about this scientific soap opera, it is no laughing matter. Behe informs the readers in his review that Dr. H.B.D. Kettlewell, "after a series of professional and personal setbacks," committed suicide.

Remember that our bodies as well as those of our animals carry little repair kits in the form of enzymes that "fix" mutations if they find them. They are not always successful in their vigilant search.

Many mutations in the DNA of somatic cells don't bother anyone and therefore often go unnoticed. We call those *silent mutations*. They also disappear with the death of the animal carrying them. Those in the DNA of the germ cells, however, cause permanent changes — some good, most bad

— since germ cell mutations are passed on to future generations. Interestingly, mutations can actually reverse themselves and restore genes to their original forms.

Through normal breeding practices, we as breeders cannot deliberately cause mutations. We can and do influence traits through selective breeding choices, molding animals much as an artist molds a sculpture. You will see in the next chapter how the "artist" sometimes gets carried away!

Chapter Twenty-five

BREEDING FOR EXTREMES

CHAPTER TWENTY-FIVE

Breeding for Extremes

"Why are people never content with what they have? What is the purpose of changing thesize of sheep? If they want a big sheep breed, why don't they go out and buy Suffolks?"
Carol Winchell (sheep breeder)

Why not, indeed? I swear there is a little Dr. Frankenstein hidden in all breeders. We just can't seem to leave well enough alone. There exists an almost bizarre compulsion to tinker with the functional traits of animals to fit our perceived needs and expectations.

Geneticists and other sensible souls continually warn against selecting for extremes. Many breeders don't listen, quite unaware of the damage they cause until it is too late. As a society, we admire and reward superlatives. The tallest building, the longest leap, the largest pumpkin — even the largest weed received a prize at the Ag Progress Days sponsored by Penn State University (in two categories, no less: broadleaf and grass).

Novice owners of large dogs are often inordinately proud of owning and exhibiting the biggest specimen among their competitors. It doesn't matter that their Great Dane, Scottish Deerhound, Irish Wolfhound, or Borzoi is slabsided and much too straight in front and rear angulation. Size is what counts in their book! Some go in the other direction and strive for the record of breeding the tiniest toy dog in the country. So what if all its teeth fall out by the time it's two years old and the patellas (knees) are unstable — its record will stand for quite some time. If the dachshund has a long back, then let's give it an even longer back, no matter that the unfortunate dog will need expensive back surgery before it is three years old! (Although disc disease is common in dachshunds, I must state here that some researchers dispute the validity of the claim that extremely long backs cause problems.)

I read that domestic turkeys are bred for such an abundance of breast meat that it is physically impossible for the toms (males) to naturally breed the hens (pictures of certain weightlifters come to mind, and ... oh, never mind). Only artificial insemination (A.I.) keeps the domestic birds alive and reproducing. Some dog breeds cannot whelp (give birth) without the assistance of the very humans who have caused the problems by selecting for extreme conformation that makes natural births impossible.

An alpaca breeder once proudly stated that the average size and weight of his animals had increased by 25 percent since he started his breeding program. Breeders familiar with *spider leg syndrome* in sheep may very well shudder upon hearing such a boast.

Several years ago, show judges started to award blue ribbons to the exhibitors of larger and longer-legged Suffolk sheep. Breeders began to select for these traits in their stock. Eventually, a genetic defect aptly named "spider leg syndrome" (*chondrodysplasia*) surfaced in those flocks.

The Merck Veterinary Manual (1998) describes *spider leg syndrome* lambs as having "pronounced medial deviation of the carpus and hock and are unable to stand without distress." The back shows dorsal rounding, and the structure of the skull is affected. Most lambs suffering from this disorder have to be euthanized. The Merck Manual explains this defect as being inherited through a simple autosomal recessive mechanism.

Fortunately for Suffolk sheep and their breeders, a DNA marker for this condition was found. Breeders can test their stock for carriers. Dr. Sponenberg suspects that carriers ($S\ s^p$) actually show minimal expression, making their phenotype unfortunately desirable for the show ring. That particular blue ribbon carries a price no breeder in his right mind wants to pay.

Selection pressure for large size in other livestock species also brought birthing problems, delayed maturity, and longer gestation periods.

Another consideration for some species is the fact that smaller animals handle heat stress better than their larger counterparts. Readers who doubt the validity of this statement can research the functional traits of various cattle breeds and their adaptation to climate.

A visit to a sighthound race meet or lure coursing field trial on a hot, humid day can be a real eye-opener. Denise, who races and lure courses with both large and small sighthounds, has observed that the difference in body mass between the Whippets (roughly 25 to 40 pounds) and their larger cousins, the Greyhounds (usually ranging from 50 to 90 pounds), causes the Greyhounds to be the first to suffer from heat prostration, heat stroke, and other heat-induced metabolic disorders.

I always assumed that all rabbits bred like, well, rabbits. Leslie Samson's article in an issue of *The Fiberfest* taught me differently.

According to Samson, increased fiber density in Angora rabbits translates into reproductive difficulties. She writes: "Thyroid studies of German Angoras showed their T3 and T4 levels to be up to four times higher than those of the thinner-coated, English type Angora rabbits. Simply put, in order to push the density to ever higher levels, something had to give."

Consider this, if you are a breeder of any species. Nature will not be cheated. She seeks a fine balance in all her children. If we upset that balance, we (or our animals) eventually pay the price.

The *AKC Gazette* devoted its September 1999 issue specifically to performance events. While reading various and interesting breed columns, I came across the following quotes:

"When any aspect of a dog becomes overemphasized in the show ring, and thus the whelping box, it is always to the detriment of the breed" — Virginia Antia, *Manchester Terriers*.

"Fanciers have a great responsibility to keep faith with 5,000 years of true breeding — they must shun man's natural tendency to 'improve,' which so often in dog breeding terms means to alter out of all recognition" — Rita Laven-thall Sacks, *Pharaoh Hounds*.

Breeders of dogs — as well as other species — ignore this advice at their peril!

Quite often, the absurd quest for extremes and breeding fads are intertwined. To illustrate my point, in the following chapter I'll reprint, with minor changes and additions, excerpts from an article I wrote for the 1999 Summer issue of the *Lama Letter*.

Chapter Twenty-six

FADS OR FUNCTION?

CHAPTER TWENTY-SIX

Fads or Function?

"Perhaps even worse than an aimless approach is a breeding program that chases after the fads dictated by the show ring or other people. Such a program is constantly acquiring, discarding, and thus wasting genetic resources in a futile attempt to catch up to the current leaders in the breed."
D. Phillip Sponenberg, D.V.M., Ph.D., and Carolyn J. Christman, D.V.M., Ph.D., *A Conservation Breeding Handbook*, 1995

Sometimes I wonder — would I have enjoyed fame as a model of the Flemish painter Peter Paul Rubens had I been born four hundred years ago? Today's fashion standard of whippet-thin bodies makes it difficult to find dresses that fit my unfashionably stocky physique. My body type, described by my physically delicate but bluntly outspoken younger sister as "robust peasant stock," has not been in vogue for years. This does not cause me any loss of sleep, though it did lead me to ponder the fads governing not only human traits but also those of the animals bred by us. In *Cloning*, Boston University ethicist George Annan asks, "What is a better human being?" He answers his own question by stating, "A lot of it is just fad."

As talk of a breed standard floats around the alpaca industry, it behooves us to clarify which traits clearly serve a functional purpose and which ones are fads created by clever advertising. Exactly what niche do we expect the alpaca to fill?

Before I address specific alpaca issues, let's examine how fads impact the health and ultimately the commercial value of other species. The general public is becoming more familiar with the concept of genetic and health problems in the dog world. Selective breeding based on fads in the show ring has brought physical misery to untold generations of dogs, and emptied the pockets of their owners in the form of huge veterinary bills.

Following the misguided belief that "more must be better," ignorant and uncaring breeders exaggerated traits to the point of developing stylized caricatures of formerly functional animals. Over-angulated rears and floppy pasterns crippled the German Shepherd. The Afghan Hound, once a proud and skillful hunter over rough terrain, is largely a crate dweller or couch potato, the maintenance of his nearly floor length coat only allowing brief exercise on impeccably manicured grounds. The broad heads and slim hips of several breeds make it impossible for bitches to give birth without human intervention. Virtually all of their puppies are born by Caesarean section.

Established breeders of the Border Collie were so terrified of losing their dogs' fabulous herding ability to fads in the show ring that they fought AKC acceptance of their breed tooth and nail. They lost. It might be a small comfort to them to see that the tide has turned. Dog breeders came to realize that pet owners were tired of continually carting their dogs to the vet's office. The average owner is more interested in enjoying healthy pets than show winners. In an ideal world, such winners would also be the healthiest, most functional animals.

The famous Merino sheep offer another example of fads run amok. Many Merinos at one time sported folds of skin heavy enough to remind one of a Chinese Sharpei. These folds were "created" to allow increased wool harvest per animal. However, the micron count varied considerably between the exterior and interior of these folds, and shearing could take as long as 2.5 hours for some of the larger rams. At that time, breeders were obviously more interested in promoting this fad than overall health and easy maintenance of their sheep. Barbara Platt describes in *sheep! magazine* (yes, that's how it's spelled) how "despite their heavy fleece production and their ability to forage where other breeds failed, the American or Vermont Merino had their drawbacks. The ewes were not prolific, were poor milkers, and their lambs were weak." Platt tells how one breeder "declared them to be so feeble that for the first 24 hours it was seldom possible to determine if they would live." Scary, isn't it?

Merino breeders fortunately came to their senses and most bred out the trait for the heavy folds. Apparently, an overlap on the neck remains and still poses a special challenge to a shearer. Breeders did manage to get their Merinos back on the path to hardiness and overall fitness.

Many sheep breeders at one time preferred a muffle-faced sheep (excessive fiber growth on the face). This fad may hold significance for alpaca breeders because of the discovery long ago that "wool blindness inhibits eating and mothering," as stated by well-respected sheep breeder and author Paula Simmons. Simmons informs the reader in her book *Raising Sheep the Modern Way* (1989) that Australian tests "have proven muffle-faced ewes to be less fertile and productive." In the 1970's, breeders made a concerted effort to eliminate this trait in their herds. I was told by Edie Van Valkenburg, a South Jersey breeder and judge of Jacob sheep, that the majority of present-day breeders prefer and choose open faces.

Jacob sheep are an ancient breed and were at one time in danger of becoming extinct. Thankfully, fanciers appreciate their unique look. Handspinners love the wool of these spotted beauties (photo by Edie Van Valkenburg)

The glossary of Dr. Gauly's book *Neuweltkameliden* (1997) describes a "tuco (entucado)" as "Tier mit unerwuenschtem uebermaessigem Faserwachstum im Gesicht, Ohren und Augen verdeckend" (translation: "an animal with undesirable excessive fiber growth in the face, covering ears and eyes").

Presumably a muffled face, both in sheep and alpacas, connotes a super dense fleece. Super density sounds desirable, but at what price?

Every geneticist whose work I've read advises moderation in selecting for specific traits. I marvel at the arrogance of individuals who totally ignore the combined recommendations of those with vastly superior knowledge.

Who knows how fads get started? Sometimes it takes just one aggressive individual promoting a certain trait heavily represented in his herd, and … ta da! … a fad is born. Even knowledgeable people are often too lazy to properly investigate and research the validity of claims made by these individuals. Novice breeders don't know enough to question them at all.

Since writing about muffle-faced sheep and alpacas in *The Lama Letter*, I found this pertinent quote from Rigoberto Calle Escobar. He wrote: "The laniferous formation of the head even in the cases of greater coverage always leaves a strip along the nose where the fiber is absent, a feature that avoids blinding as it occurs in sheep because of excessive fiber on the face."

Doctors Sponenberg and Christman also tell us that "when sheep were intensively selected for wool coverage, for example, the result was expected to be heavy fleeces. In actuality, this selection produced 'wool blind' sheep with poor survivability."

Before taking on the responsibility of breeding, any person worth his or her salt will question the purpose and function of traits in their chosen species or breed. If you don't know why your dog's breed standard calls for hare feet, well sprung ribs or a harsh, wiry coat, make it your business to find out. Knowledgeable, ethical breeders will not resent or evade your questions.

Analyze the meaning and purpose of advertisements in animal magazines. An alpaca breeder praised his animals for having short muzzles ("the projecting part of the head of a dog, horse, etc." — Webster's Dictionary). Short in relation to what? Shorter than those of the average alpacas grazing in North American pastures? Should muzzles be short, and if so, why? Is a very short muzzle a functional trait for a grazing animal? Does it affect bite? Will selecting for extremely short muzzles impact ease or difficulty at birthing? Do correlations of head shape to other body parts exist in alpacas as they do for sheep and other species?

Breeders of Burmese cats who selected for extremely short heads lived to regret that decision. It resulted in severe neurological problems. In the *AKC Gazette* (March, 2000), Kim Campbell Thornton tells us in her article about Pomeranians: "The breed's shortened muzzle has made it more difficult for the mouth to accommodate all the teeth." At the other extreme, dog breeders who selected for extra incisors (resulting in the broader heads they desired) were informed by geneticists that deviations from the norm in regard to teeth were tied to abnormal calcification of the bones in such cases.

In Cocker Spaniels, researchers found a mild degree of hydro-encephalus. Breeders had created this brain defect by selecting for a certain "domey" head shape deemed desirable in the show ring.

Interested readers may wish to study the concept of *neoteny*, the retention of juvenile features in an adult animal. In many species, domestication produced a more juvenile head shape that features a shortened muzzle. Denise reminded me of research suggesting that both animals and humans are more drawn to, and take better care of, "cute" juvenile-appearing animals. The authors of "Behavioral Genetics and Animal Science" (from *Genetics and the Behavior of Domestic Animals*) refer to several researchers who have done pertinent studies in that area.

In the *AKC Gazette* (January, 2001), Golden Retriever breeder Jeffrey G. Pepper wrote: "The muzzle should be about the same length from stop to nose as the skull from stop to occiput, never substantially less, even if a short, bear-like muzzle looks cute to you."

An elongated head makes functional sense for those animals that have to either see predators or prey at a far distance. Such head shapes are advantageous, for example, for the vicuña (prey) and all sighthound breeds (predators). The heads of horses also became longer as the species adapted to life on the open plains. With such a head shape, horses "evolved the ability to see with binocular vision as well as monocular vision" (James C. Heird, et al).

Alpaca breeders who believe that a short muzzle always correlates to an absence of llama genes may want to re-visit Mendel's Law of Independent Assortment. Breeders of other species have proven how easy it is to change head shape on animals without resorting to crossbreeding. Aside from that, one specific physical feature does not determine the entire genetic make-up of an animal.

In an earlier chapter, I briefly discussed an important discovery made in South America. Jane Wheeler, an American archaeozoologist, proved with DNA testing that roughly 90 percent of all tested alpacas were llama-alpaca hybrids. Breeders should remember this Wheeler quote when they read certain advertising claims: "The other thing we discovered is that it's not possible to tell whether an alpaca or a llama is a purebred by looking at it. It's necessary to do DNA testing to certify purity" (*Discovery Magazine*, April 2001, pp 58-65).

In *Animal Breeding and Production of American Camelids*, Rigoberto Calle Escobar tells us that alpaca's legs "are thin and agile with strong musculature ..." An article in a North American publication presents thin legs as an undesirable trait. Why? Another author claimed that heavy bone presented a desirable trait in the harsh Peruvian climate.

Susan Tyler, an Australian scholar of medieval Japanese culture and fellow alpaca breeder, questioned this last statement and wrote in personal correspondence: "I cannot see why they should need thick leg bones for coping with the severe environment of the Andes, when vicuñas do not seem to be heavy boned and yet cope just fine ..." Vicuñas are not domesticated — for thousands of years they've procreated based on the old rule of survival of the fittest.

Although I have not had the pleasure of observing a live vicuña, the numerous photos I've seen of these delightful camelids all show animals with comparatively delicate leg bones. Their muzzles cannot be described as short by any stretch of the imagination. Research has proven the vicuña to be the wild ancestor of the alpaca. Draw your own conclusions.

Dr. Harry Preston, praised by Professor Escobar as "a pioneer in scientific matters dealing with the health of the South American Camelid" in Peru, also described the legs of alpacas as "thin and agile with a strong musculature."

Alpacas are natural pacers. Pacing is an extremely efficient mode of locomotion for grazers. It cannot be sustained for a long time by animals with heavy bone accompanied by wide fronts. Of course, selecting <u>against</u> super heavy bone <u>does not</u> translate into selecting <u>for</u> weedy animals lacking substance. Moderation and common sense are the key words here.

Lynd Blatchford, a llama breeder from Maine, gave me a good laugh with his *Observations from the Pasture* (*GALA Newsletter*, August 2001). Among other tidbits, Blatchford confessed to a "new awareness." He wrote: "My epiphany came when I saw several very large, heavy-

A BREEDER'S GUIDE TO GENETICS
Relax, It's Not Rocket Science

wooled, heavy-boned llamas which could best be described as oxen with banana ears." He questioned whether these animals were "up to the task of being llamas?" In South America, llamas were and still are typically bred to serve as pack animals, and are expected to travel over great distances. Would such conformation be functional?

Sighthound breeders have taken issue for years with those who try to add heavy bone to the phenotype of their hounds. While possibly appropriate for a mastiff or a Saint Bernard, it is counterproductive to the functional gait (double suspension gallop) of a Borzoi or a Greyhound, for example.

Some camelid breeders prefer heavy fiber coverage on the legs of their animals. While I admit that it looks "cute," is it more functional than sparsely covered limbs? Should "cuteness" guide our breeding selections? Leg fiber has little commercial value (too coarse) and is a nuisance to shear. In my trusty sheep book, author Paula Simmons advises breeders to avoid the following: "Wool going too far down on legs. It is more trouble to shear."

Simmons modified her stance somewhat in *Storey's Guide to Raising Sheep*. She allows that in extreme climates "wool on the legs and head, like socks and a hat on people, help the sheep maintain body temperature."

It should be recognized that a trait may be functional in one environment and just the opposite in a different one.

Afghan Hound breeders can, or should, identify with the "overdone fleece" situation. Georgie Guthrie (*AKC Gazette,* August, 1999), admonished dog show judges for not recognizing that short hair on the lower legs of some Afghans (Persian cuffs) is entirely appropriate for that breed. What's more, it is specifically mentioned as being permissible in the official AKC breed standard. And why shouldn't it be? No one can convince me that an Afghan Hound with heavily coated pasterns is a more functional animal in the field!

The sad fact is that many breeders only give lip service to the concept of striving to select for functional animals, without fully understanding the reality of the situation. The even sadder truth is that functional conformation is often not what puts the blue first-place show ribbons in the hands of owners and breeders. In the *AKC Gazette* (August,

2000), Scottish Deerhound breeder Joan Shagan candidly shared with readers the story of Andy. She describes this male out of her first litter as "one of the most handsome puppies." Unfortunately, Andy "was also very unsound coming and going." When Shagan showed him to help build points for other dogs, this structurally unsound Deerhound became the first male in the litter to finish his champion title. His breeder kept Andy as a pet but did not use him in her breeding program.

I found an interesting article on pasture lambing in the 2003 *Premier* supply catalog. Under the sub-topic of flock genetics, I read: "Large ewes with small stomachs, though ideal for the showring, cannot consume enough dry matter to produce enough milk to support two young lambs…"

Another example is the Churro sheep, owned by the Navajo Indian tribe. The Churros were substantially weakened by the introduction of "improved" breeding stock not suitable for their environment. We must acknowledge the fact that extensive human intervention, genetic "tinkering," and programs to "improve" a breed or species can be harmful to the animals.

Novices do not always understand why experienced, knowledgeable breeders select for or against certain traits. Before you judge these decisions to be completely arbitrary and possibly frivolous, conduct some research of your own. You might be surprised at what you uncover.

For example, one author I read described the long horns of the African Ankole Watusi cattle as "whimsical." This is not so! Carolyn J. Christman, et al, explains in *A Rare Breeds Album* (1997) "the horns are part of adaption to a hot climate by allowing dispersal of excess body heat."

Without ignoring breed standards and common sense, breeders can apply their own interpretations and tastes, creating the healthy diversity necessary for the ultimate survival of the species. I realize that I've made this point before — but it's an important one and bears repeating. No matter what your own personal vision or dream is, before you decide to embrace a trend, ask yourself: "Is it fad or function?"

Chapter Twenty-seven

PEDIGREE POWER — PEDIGREE PERILS

CHAPTER TWENTY-SEVEN

Pedigree Power — Pedigree Perils

"Pedigree: a table, chart, diagram, or list of an animal's ancestors, used in genetics in the analysis of Mendelian inheritance, and in the prediction of productivity and breed quality in the offspring."
Blood & Studdert, *Baillière's Comprehensive Veterinary Dictionary*, 1988

The study of species other than those we own or breed has much to offer to anyone eager for knowledge. I intend to continue to learn from the mistakes, as well as the triumphs, experienced by the breeders of horses, sheep, mice, cats — they all have something to teach us.

My initial interest in fiber-bearing animals other than alpacas was sparked by the wonderfully unpretentious magazine *The Fiberfest*. The publication unfortunately ceased with the untimely death of publisher and fellow alpaca breeder, Sue Drummond. Filled with practical information, *The Fiberfest* offered sensible advice on owning and breeding fiber-bearing animals. An astute reader could often discover parallels of pertinent knowledge that bridged the gap between various species. At the very least, he or she was motivated to question conventional breeding practices.

While reading *The Bull* (Allen Fraser, 1972), I learned that pedigrees are nothing new in the livestock world. In tiny Austria, the herdbook of the dual-purpose Pinzgauer cattle is over 1,400 years old (no, this is not a misprint). Centuries after the opening of that book, an adventurous English livestock breeder used a system of inbreeding to create more homozygous gene pools in the various species on his farm. The contemporaries of Robert Blakewell (1725-1795) were appalled at the poor results when they viewed his Berkshire pigs and Longhorn cattle. However, they eagerly sought Blakewell's advice after he succeeded admirably with his Longwool Leicester sheep. This illustrates a fascinating enigma: some species or specific breeds within a species do not suffer the negative impact of inbreeding as quickly as others do.

In England and Scotland, pedigree mania eventually overtook breeding programs. Looking back, the *British Agricultural Bulletin* (1950,

Vol. III, No. 3) called it "as much cult as commerce." Breeders of Shorthorn cattle eventually became obsessed with real and perceived pedigree power. Pedigree alone, rather than a healthy balance of lineage and the actual living animal, drove the industry. Allen Fraser wrote: "Nevertheless, pedigree in Shorthorn cattle became almost a fetish for a time, so much so that very poor cattle with impressive pedigrees sometimes sold for high prices..."

A pedigree is meaningless to breeders who are not familiar with the animals represented by the names. It can be a useful tool for those who take the time and trouble to research it. What did sire and dam look like? Were they "typey" in addition to having good conformation? What were their temperaments like, and how were those possibly influenced by environment? How about genetic defects carried in their lines? Were they healthy? Were the dams on their respective sides of the pedigree good mothers? Did they give birth easily and provide plenty of milk and good nurturing to their offspring?

Denise advised me that our illustrator, Maryann Conran, has been collecting photos of the dogs in her Borzois' pedigrees. She also records as many personal observations as possible from folks who actually owned or bred those dogs — for precisely the reasons I mentioned in the last paragraph.

Experienced breeders know that even the best "paper matings" (breeding pedigree to pedigree without taking the animals into consideration) don't always produce the desired results. Recessive genes, the occasional mutation, and environmental influences can and do affect the outcome of a breeding. Often there is no ready explanation for the occasionally disastrous outcomes — sarcastically called "paper tigers" by old-time breeders — for these seemingly perfect genetic unions.

In *The Animals in My Life,* author Grant Kendall exclaims: "Always Angie! The pedigree of the century belonged to this ... this ... pony? This lop-eared, dead-headed, doe-eyed pony?"

Fate can also bestow unexplained success. A good example is Amos Cruickshank's purchase of the bull "Lancaster Comet,"as told to us by Fraser. Amos, residing at Siltyton Farm in Scotland, bought the bull, sight unseen, from a well-respected Shorthorn breeder. He was so disgusted at his first glimpse of Comet that he banished the animal to a remote

A BREEDER'S GUIDE TO GENETICS
Relax, It's Not Rocket Science

subsidiary farm. The rejected bull was bred to a cow named "Virtue of Plantagenet." Comet died shortly thereafter. It's a good thing Virtue didn't keep her virtue, because Shorthorn history was made with that breeding. A bull calf, "Champion of England," was born in 1859. One hundred years later, Commander T. B. Marston described the union as "founding a dynasty." Luck, fate — the Gods of Genetic Success sometimes smile upon such haphazard breedings. Fraser's book is fascinating reading.

Readers may be familiar with the story of the Morgan horse, a breed founded by a nondescript little colt named "Figure," taken in payment of a debt by Justin Morgan, a Vermont schoolteacher (circa 1795). Maturing into a smallish but powerful and incredibly versatile animal, this prepotent stallion reproduced himself so exactly that an entire breed of strong, intelligent, and multi-talented horses resulted, each stamped with the type of their progenitor.

Reading and interpreting a pedigree is not that complicated in well-established species or breeds, once one has totally familiarized oneself with many individual animals. Of the fourteen dogs listed in my recent litter's three-generation pedigree, I know eleven very well. Two live in my household, the other nine I observed racing or lure coursing over many weekends for a number of years. Their conformation as well as their enthusiasm and ability to handle pressure are imprinted on my mind. I've also accumulated valuable knowledge of their littermates, and of the progeny produced by those animals thatwere bred to others not listed in the pedigree of my puppies. There is, unfortunately, no shortcut to acquiring such a library of genetic traits.

Breeders of alpacas and other recently imported species without documented lineage have a tougher row to hoe. Such animals are named by their new owners, but that in itself provides no real genetic information to a breeder. I realize we've made this point before, but it bears repeating.

To illustrate this particular predicament, let's imagine the llama herds of breeders A and B.

"A" owns Harry and Betty. "B" owns Joe and Cindy. Harry and Joe are recent imports. Unbeknown to their new owners, they are also full brothers. Let's call their sire and dam back in South America Bill and Beth. Breeder "A" mates Harry to Betty, resulting in the birth of a female cria he names Linda. Since Mr. "A" does not want to risk the possible

Ingrid Wood and Denise Como

ramifications of inbreeding, he decides against breeding Linda back to her sire, Harry. Instead, he contracts with Mr. "B" to use Joe as a stud. Let's see what happens:

```
                    ┌ Joe    ┌ Bill
                    │        └ Beth
Oh Boy I ───────────┤
                    │        ┌ Harry ┌ Bill
                    │        │       └ Beth
                    └ Linda ─┤
                             └ Betty
```

No single name appears more than once in the 2-generation <u>North American</u> pedigree. Remember, the pedigree doesn't list Bill and Beth as Harry's sire and dam — the breeder has no information on the animals' identity in the third generation. What would the pedigree look like had Mr. "A" chosen to breed his female back to her sire?

```
                    ┌ Harry ┌ Bill
                    │       └ Beth
Oh Boy II ──────────┤
                    │        ┌ Harry ┌ *Bill*
                    │        │       └ *Beth*
                    └ Linda ─┤
                             └ Betty
```

Oh boy, indeed!

Denise noticed this same phenomenon when she did extended research into Borzoi pedigrees. Many pedigrees looked like outcrosses until she reached back past 6 or 7 generations, and farther — it became apparent that the "different" bloodlines actually began to narrow down to a small gene pool, consisting of very few dogs. Breeders must take into consideration that littermates may not carry the same kennel names, and that there were more father-daughter, mother-son, brother-sister breedings done in the early years than most folks are aware of. As the old doggerel goes, "Julie O'Grady and the Colonel's Lady are sisters under the skin."

By the way, generations are customarily counted back from the sire and dam — so a 4-generation pedigree would show sire and dam, grandparents, great-grandparents, and great-great-grandparents.

Without a doubt, pedigrees enhance name recognition enjoyed by well-advertised studs. Often owners will breed to a famous stud for the sole reason that he is famous. That stud may be great, but if he doesn't possess the traits your female needs, you will not be happy with the results. Little-known animals without "credentials" but of excellent quality can be standing in small herds or kennels. They can serve your purpose just as well, if not better, than the well-advertised and pricey stud. It does take diligent detective work to find such animals and admittedly can be a difficult process for a novice. The use of such unpublicized studs is absolutely crucial to conserve and expand genetic diversity, especially in species or breeds where registries are closed to imports, and animals are not bred for human consumption.

Genetic drift, the random loss of genetic variants from a population, occurs as a matter of course in wild populations, especially small ones. Vigorously selecting only from the genetic material of a severely limited number of studs is not drift — it's a genetic avalanche burying valuable genes in its path.

Do pedigrees greatly benefit a pet or livestock breeding program? They don't in the hands of unscrupulous people who use them to "razzle dazzle" unknowing buyers. They don't in the hands of novices that think memorizing and rattling off chains of names convey intimate knowledge of the breed. They do in the hands of ethical and informed breeders who study them for the express purpose of making truly educated decisions with the goal of improving their stock. To serve any real purpose, pedigrees must be used with an understanding of the animals involved.

A long pedigree in and of itself is a meaningless jumble of names. It does not necessarily represent quality. A "famous" pedigree doesn't always translate into the highest value, either. To quote dairy cow expert Elwood M. Juergenson (*Approved Practices in Dairying*, 1977): "It is well to consider that the fanciest barns, highest priced livestock, and the glib-tongued seller do not necessarily represent the best stock available for the money." These are timeless words of wisdom!

Neither the pedigree nor the phenotype of an animal always tells the full story. That brings us to consider another important aspect of any serious breeding program — *nurture helping nature.*

Chapter Twenty-eight

NURTURE HELPING NATURE

CHAPTER TWENTY-EIGHT

Nurture Helping Nature

"The greatest failing common to all branches of livestock fanciers is the tendency to keep too many animals and what should be a hobby to occupy one's leisure moments tends to become a task and a responsibility."
W. MacKintosh Kerr, M.B., Ch.B., *Colour Inheritance in Fancy Mice* (no date given)

A while back, I enjoyed a most pleasant visit to the llamas of Morning Star Ranch. Located just beyond the quaintly beautiful village of Crosswicks, New Jersey, Judy Morgenstern's ranch is home to twenty llamas of all ages, sizes, and colors. All of my days should be filled with such tranquility.

Disregarding the scorching heat, Judy graciously introduced me to her doe-eyed charges. I had always wondered how the tiny Judy handled such large animals. Now I know. Judy uses gentle care and calm gestures — not to mention an obvious abundance of love, affection, and meticulous attention to detail. It shows!

The silly assumption that all llamas spit constantly at everything (people included) was not given any credibility whatsoever at Morning Star Ranch. With the relentless sun beating down on pastures and barns and the rumbling promise of an afternoon thunderstorm in the air, the llamas had every right to be irritable and act out-of-sorts.

Nothing could have been further from the truth. Not only didn't "spit happen," but all the animals reacted toward me, a total stranger, with the same charming inquisitiveness. I received lotsof airy kisses, and not one animal objected to being petted.

Is Judy lucky? Do all her animals just happen to have nice temperaments? I don't think so. Her llamas' daily exposure to and interaction with kind and patient caretakers fosters a trusting attitude towards two-leggeds in general. Judy stressed the fact that she vigorously selects for good temperament. Animals are no different than people, in that genetic contributions and environment <u>combine</u> to form the individual. A poorly fed, poorly housed, and poorly treated animal will never reach its full, genetically programmed potential.

At one time, many scientists were of the opinion that only environment formed the temperament and behavior patterns of an animal. That is no longer the case. James C. Heird and Mark J. Deering, for example, discuss research published in *Science* (1990) by R. Plomin. They write in *Genetics and the Behavior of Domestic Animals:* "It is generally concluded that behavior is determined by a complex interaction between the genes that an individual receives from its parents and the environment. No single gene controls any single behavior."

Horses, camelids, and sheep (among other prey animals) possess a strong flight instinct. Owners and breeders will be interested in a fascinating study, cited by Dr. Temple Grandin and Mark J. Deering, in which researchers found that "the most reactive pigs were also the ones that were the most attracted to novelty." Alpacas are certainly the nosiest, most inquisitive creatures I've ever come across. Denise said the same about her horses.

I highly recommend the above-mentioned book. Fifteen scientists, among them editor Dr. Temple Grandin of Colorado State University, present a wealth of fascinating information as well as extensive references in its eleven chapters. After reading it, you should fully appreciate the enormous complexity of an animal's genome. Trust me, the purchase price will be money well spent.

Without diminishing the role that genes play in shaping an animal's temperament, it is still safe to say that nutrition, handling, and other outside factors play important roles in how an animal reacts to and interacts with its surroundings.

It's not always easy to determine the reasons for an animal's behavior.

Is the alpaca stud's disposition nasty because it is inherited, or was he mishandled in his formative youth? A bottle-fed and fussed-over llama or alpaca can act pushy and even aggressive with people — would the same animal be reasonable and manageable if it did <u>not</u> imprint primarily on people as a cria? This may be a difficult concept to appreciate for those of us who also raise dogs — and who give our puppies every opportunity to bond with humans. The selection criteria for puppies (predators) do not apply to camelids (prey animals).

Beware of overly friendly, fearless camelid babies — they often grow up to be obnoxious, aggressive adults. <u>Do not</u> buy unweaned youngsters with bottles hanging around their necks at auctions or in sales barns. Trust me on this! Bottle-fed crias can grow up to be well-mannered adults <u>if</u> the human caretaker approaches the situation with knowledge and common sense. Alpaca orphans need to be socialized and taught the ropes by strict and stern alpaca "aunts" — not indulgent two-legged foster mammas.

Other variables enter into the picture. It is always interesting to observe subtle personality transformations in my little group of Whippets as well as my herd of alpacas when the population changes. Gentle, ladylike Hannah took over the daunting task of disciplining eight lively (a pleasant euphemism for hyper-active) Whippet puppies when it became obvious that their mother was too tolerant of their nonsense and foolishness. We applauded Hannah when she firmly put her brazen nephew Sonic in his proper place among the Whippet pack. Usually quite unassuming, Hannah fairly shined and blossomed under the "burden" of her self-appointed responsibilities.

Dogs also love to nurture their humans. The Whippets that belong to the two McDonald clan households (friends of mine) know the immense pleasure it gives their owners to feed them. Instead of contentedly lolling on the couch or stalking pesky squirrels in the yard, the Whippets — self-sacrificing little creatures that they are — make sure that at mealtimes they are positioned where life's duty commands their presence: underneath the well-supplied McDonald table!

All joking aside, dogs enhance our physical and emotional wellbeing in truly astounding ways. The *AKC Gazette* devoted their entire October 2000 issue to exploring and describing dogs' diverse services to mankind. From Schutzhund work to visiting the elderly to bringing autistic children out of their shells to detecting certain forms of cancer — it is hard to imagine a world without the nurturing our canine friends provide to us.

Of course, we are forever being told by authors like Stephen Budiansky (*The Truth About Dogs*, 2000) that, far from being protective and nurturing, dogs are scheming manipulators who "play us like accordians." Let's be honest — Budiansky is right. They are and they do! The fact is, we don't care!

Millions of people would emphatically agree with Denise: "I recently 'enjoyed' an unscheduled Hospital Holiday. Since I was being treated aggressively with anti-coagulants — heparin and warfarin — an imposing trio of cardiologists began to say something silly about the exuberant attentions of the dogs at home possibly causing me to suffer dangerous bruising or bleeding, and how the dogs should really... I stopped them before they finished. <u>The dogs were not leaving!</u> That was the end of <u>that</u> discussion. I told my doctors that the best therapy I could have was to be surrounded by my hounds. I was proved correct. I was not injured in any way, and my dogs provided me with companionship and emotional support."

Animals enhance our well-being by their sheer presence alone.

Watching my alpacas graze calms me after a hectic day. Their frolics at dusk make me smile. A friend who doesn't own or particularly even like animals will sit with me in the alpaca pastures and watch them for hours. It relaxes her.

We began in the alpaca business with Soft Breeze, our foundation female, and a wonderfully goofy gelding named Harley D. Breeze had impressed us with, among other attributes, her calm demeanor. Harley we bought because, well, because he was Harley (if your husband's legal name is Harley D., and you meet an alpaca with the same name, do you really have any choice but to buy him?). Both animals became much more trusting as time went on, but I am convinced that the birth of Breeze's first two crias made the real difference. Alpacas are herd animals and do not feel comfortable outside of such a setting. Two wasn't enough of a herd. Being one of four greatly enhanced the feeling of confidence and security in the two older ones. Our approach to their care had not changed, but the dynamics of their environment obviously did.

My alpaca herd offers another example of an older animal nurturing and shaping the behavior of others. Though initially somewhat shy with two-leggeds, Breeze took firm charge of her new home right from the beginning. To my great delight, she led Harley D. and then her crias to take care of their toilet business out on the pasture. Her newborns follow her faithfully, sometimes despite driving rainstorms. Even a cooling barn fan running during burning hot summer days will not entice "Mrs. Clean" to allow her charges to dirty their "house." (A good friend warned me that my inordinate thrill over this feat is a sign of a withering intellect. She, however, has never mucked out a barn and thus is not qualified to pass judgement). The nurturing that animal moms give their babies is as important, and maybe more so, as the nurturing you provide as a breeder. A good mother is worth her weight in gold in any species.

Breeders are strongly advised to carefully scrutinize the homes and farms they send their animals to. You can sell the most promising youngster your breeding program has ever produced, but if the new owners don't take proper care of him, he will not grow into the outstanding specimen you had envisioned when that animal was an infant. Many breeders of good animals have been heartsick when confronted with the horrible results of such an ill-chosen placement.

Luckily, there is the other side of the coin. I certainly have been blessed with wonderful buyers whose excellent care of and commitment to their animals has been a continuous joy to me. However, I also turn buyers away when I don't feel comfortable with them and, in some cases, talk people out of buying altogether.

Rarely do novices realize the amount of work and dedication required to become successful and to excel in breeding endeavors. I am *not* talking about financial success or points and ribbons acquired in the show ring. Owners whose animals are superior specimens often start out with above average stock, but they also leave no stone unturned and no detail unattended to while working towards their goals.

In her handbook *So, You Want to Run Your Sighthound* (1996) and in her new reference book *Sighthounds Afield* (2003), Denise offers advice to sighthound owners who are unfamiliar with competition in performance events: nature can't do it alone. Owners must nurture their canine athletes properly and wisely to achieve maximum results. Over the years, I've watched sadly as such advice was ignored by novices and old-timers alike, and hounds with promising futures were permanently ruined as a result.

As I already emphasized, you must consider environmental influences when making your breeding decisions. These choices can be tough. Is the dog's coat poor because it has a thyroid deficiency, or is it because its diet is lacking important fats? Is the Borzoi puppies' sire built like an "angelfish" (tall, narrow, roachy topline, no spring of rib — Denise taught me this expression) because of genetic reasons, or because he was confined for long periods of time as a puppy?

Unscrupulous breeders will sometimes purposely withhold energy-producing feeds. Their animals appear docile and friendly, when in reality they are simply lethargic due to poor nutrition and possibly heavy parasite loads. Years ago, I witnessed the dramatic effects of a change in nutritional levels on my own dogs. Whippets have a high metabolism and require a diet sufficient in protein and fat. I normally feed a high-performance diet all year 'round. An acquaintance voiced the opinion that such high levels of protein were unnecessary and even harmful during the off-season, when no racing or field trials were taking place. Stupidly, and without conducting my own research, I switched to another brand of feed for the winter. Within several weeks, my usually energetic and playful little sighthounds had turned into virtual zombies. When I had to literally drag them off the couch to accompany me on our nightly two-mile walks, I knew it was time to pitch the other dog food where it belonged — in the garbage.

Denise cautions readers in her performance books that different breeds and even individual hounds vary in how they metabolize food. How true!

Dr. Norm Evans, D.V.M., a highly respected veterinarian with extensive camelid experience, feels that 80 percent of all alpaca medical problems are nutrition related. Perversely, poor nutrition decreases the fiber's *micron* count (*micron*: a metric unit equal to 1/1000 of a millimeter), a most desirable result. Dr. Evans writes: "The industry's strong emphasis on fiber and its micron size has led some breeders to, in my opinion, malnourish their animals" (*Optimizing Production with Proper Nutrition*, AOBA Conference 1999).

I imagine the same concept applies to other fiber-bearing animals. Boasts of extremely low micron counts should therefore send warning signals to buyers. Nutritional deficiencies have long-ranging impact on fiber quality, fertility, skin health, and ability of mothers to nurse their young.

Environmental impact, good husbandry practices, and genetic mechanisms are closely connected and intertwined. The final product, the adult animal, should always be viewed and judged while keeping that connection in mind. All components must harmoniously complement each other if we wish to breed physically as well as mentally fit and functional animals.

Chapter Twenty-nine

AN INTRODUCTION TO THE INHERITANCE OF COLOR GENES

CHAPTER TWENTY-NINE

An Introduction to the Inheritance of Color Genes

"The color in hair, skin, and eyes are caused by the presence of melanin. Melanin is deposited in the hair shafts in the form of microscopic granules which vary in shape, size and arrangement, giving a variety of colors."
Orca Starbuck and David Thomas (from their personal web site)

If anyone had told me several years ago that I'd be writing about the inheritance of color genes in animals I would have told them they were certifiably nuts — definitely out of their minds. I initially approached the whole subject with regular good cheer and even more good will. Not for long! Soon my eyes glazed over, my brain shut down, and the book I purchased to study the subject was flung ungently into a corner.

This stuff was much too difficult! I really didn't need to know about color genes anyway, I consoled myself. All colors are acceptable in the Whippet breed, so why was I driving myself crazy trying to understand this "alphabet soup"? Periodically I'd consult my friend Barbara Ewing, a Borzoi breeder, who did manage to drum the concept of two basic pigments for mammals into my befuddled head, but that was the extent of my intellectual forays into that area.

What changed? When we bought our first alpaca female, a black with an extensive tuxedo pattern, she had been bred to a rose-grey stud (more about the unfortunate choice of nomenclature for this color pattern later). The breeder conscientiously warned me that this combination often produced phenotypical "cremellos," white animals with blue eyes.

After weighing all the options, I decided to take a chance. When several months later the blue-eyed Kalita was born, I became more intrigued with the whole business of color. I investigated again, this time with more patience and persistence (age does that for people). And no, I didn't regret my purchase. Breeze has been a wonderful animal in every respect. Her second cria, by the way, turned out to be a beautiful reddish brown with black trim, her third and fourth are black — all three have dark eyes. And Kalita, her firstborn, prompted me to study color inheritance. That led to more extensive study of genetics in general, which led to published articles and finally to the writing of this book. Kalita should probably be listed as a co-author.

Kalita's own babies, born while this book was still in progress, are both coal black with small white markings. Several visitors expressed surprise that a blue-eyed white mom had black babies. No surprise there! Unlike in dogs, where black is dominant to red, the color is carried as a recessive to red by alpacas (as it is by most other mammalian species). The phenotypically white but genetically "colored" Kalita carries at least one allele coding for black which she inherited from her own mom. Kalita, as you will learn, is positively not dominant white.

What complicates the subject of color inheritance so much? It starts with nomenclature. Sometimes colors are given different names, depending on the species or the breed. This can be very confusing when you're initially wading through various texts, trying to gain a better understanding.

Names chosen for colors by various registries can be unintentionally misleading to newcomers. "We bought a rose-grey. I just love grey," gushed a new alpaca breeder excitedly on her visit to Stormwind Farm. I didn't have the heart to tell her that her rose-grey female was not grey at all, but a red (brown) animal with white fiber evenly interspersed throughout the fleece — a roan! The breeder she bought this animal from had obviously not disclosed this little detail.

Dr. Dale Graham, a pioneer in the field of camelid color genetics, posted on her Internet web site (http://www.llamaweb.com/About/Colors.html) that she does not even contemplate the possibility of there being any true "grey" llamas. She claims that the greying gene does not exist in llamas and assigns all "greys" to the roan category, roan being either red (brown)-white or black-white. I bet that's where many of the "grey" alpacas belong as well. They are beautiful, regardless of the genetic source of their color.

The industry's choice of nomenclature for the camelid "greys" creates somewhat of a problem. The surfacing of any true greys would complicate things even more. Official color designations are best left to people with extensive knowledge of genetic mechanisms in this specialized area.

Alpaca breeders also lump *tuxedo* and *piebald* patterns together. That's not such a good idea since the patterns are believed to behave in different ways — genetically speaking. I once turned down an alpaca stud I liked very much because I had not learned yet to differentiate between the two patterns and their separate mechanisms. Ignorance is not always bliss!

At times color can be difficult to determine when the extreme expression of one color gene visually mimics that of another. Cindy Schmidt, a long-time Whippet breeder, relayed an interesting example to me. I have mentioned several times that black is the dominant color in Whippets. Cindy told me that years ago a breeder reported a black puppy being born from two red brindle parents. This is genetically impossible, but it had to be admitted that this particular puppy was indeed black. All possibilities involving a black stud were ruled out.

"Then we met at a show on a very sunny, bright day," Cindy explained. "As the sun's rays hit this bitch's coat just right, the mystery was solved. The bitch was a brindle, with a base color so dark it blended together with the brindle pattern, which made her appear to be black." Rainy and Lightning, the two brindles born to my bitch Stormy, looked almost black when they were born. The dark base color gradually faded to red. By two months of age, only the black masks and the dark, brindle striping remained.

Denise had a male Borzoi, who at 6 days of age looked like a Irish-marked black dog (with a white blaze, a full white collar and "apron," white legs, white tail tip, and a white patch over his hips). As he matured, he turned dark red, with complete brindling clearly visible over all the colored areas. He had the "widow's peak" mask like an Alaskan Malamute, Agouti "banded" hairs, and a giant mantle of dark sabling covered most of his upper body. If you parted his hair, the striping was apparent, and his skin under the non-white hairs was very dark. Depending on the color of the bitch he was bred to, Beckett produced a veritable rainbow of colors and markings — from nearly all white (extreme white spotting), to self-blacks with tan or brindle trim, including grey-brindles and an Irish-marked red dog with "banded" hairs. As he aged, Beckett developed white hairs evenly interspersed throughout his entire coat. He was a true genetic smorgasbord!

It is not unusual for certain breeds of horses to change color over several years. The seasons and level of nutrition also play significant roles.

Breeders of any species may be mistaken about the true identity of a sire. Occasionally outright fraud takes place. Erroneous information then upsets the entire color "apple cart," if pedigrees based on it are later used to determine inheritance patterns.

At times, even geneticists cannot track the exact mode of inheritance and will freely admit it. There are, of course, the usual disagreements among individual breeders and authors. Only people new to the fields of genetics and breeding get bent out of shape over outright clashes of opinion. Once immersed in the subject, you learn to accept these differences as par for the course. You might even embrace those hot debates!

Although the inheritance of color and pattern genes vary from species to species, there exists a basic "blueprint." You may want to call it a "system."

Let's start out with an excellent visual aid to help our understanding. Dr. Sponenberg uses the word *switches* to explain the choices between alleles at each locus.

This led me in turn to visualize the following scenario:

Imagine that you are sitting in front of a console engineered to produce color. A picture of a dog is displayed on top of the console — and he is absolutely devoid of all color and markings. It is your job to flick the appropriate switches to fill in the color. Two banks of switches are labeled with letters, and you must follow a specific order in turning them on.

Each switch on the dam's side of the console has a corresponding "partner" on the sire's side. As you flick switches, the picture on top of the console slowly fills in with color and pattern, which change periodically according to the additions you make. Eventually the offspring's final color phenotype emerges. It's important to understand that genes don't "blend" together. They remain separate entities affecting each other to various degrees.

The numerous color phenotypes enjoyed by many species serve as an excellent example of polygenic inheritance. It takes more than one gene to produce, for example, a grey dog or a bay horse. You will see how a series of loci must work together to achieve colors and patterns. Allelic combinations at various loci that result in specific phenotypes will separate in future generations (*Principal of Segregation* and *Mendel's Law of Independent Assortment*). These are crucial concepts to grasp.

Later, we'll use my Whippet Rainy, a red brindle with a black mask, as an example to clarify color inheritance. I had originally planned — and would enjoy — using a llama or alpaca instead of a dog to demonstrate how we can "construct" a genotype with corresponding phenotype, but there remains too much speculation concerning the inheritance of color in camelids.

Before we advance to manning our color console, let us continue to cover the basics and discuss individual loci in more detail.

An important concept you need to understand initially is that all color in hair or fiber is the result of melanin granules that vary in size and shape. They produce two pigments called *phaeomelanin* (also spelled *pheomelanin*) and *eumelanin*. Phaeomelanin colors hair/fiber red, eumelanin colors them black. "Now, wait one minute!" you exclaim. "There are more colors than just red and black!" Not really! It just looks that way because of the other switches you can flick on that console.

Geneticists advise us to think of any animal, even a white one, as genetically either black or red. This concept will become clear as you read about color loci.

Believe it or not, all the various shades are derived from the two basic pigments. As a rule of thumb, dark pigment is dominant over lack of pigment, although there are several exceptions. At certain loci, dominant genes code for lack or reduction of pigment. For example, species such as the horse, cat, and sheep have genes that produce dominant white coats and fleeces. I mentioned in a previous chapter that such horses have dark eyes. The skin is pink, including the rims around the eyes. Dr. Dale Graham denies the existence of dominant white in llamas.

Allow me to stress the important concept Dr. Graham based her opinion on: you cannot determine the genetic color make-up of a fiber-bearing animal by judging only the fleece color. "Because the dominant white gene suppresses pigmentation in the skin, this means that there should be no darker pigmented places on the skin" (from personal communication). If a llama's fleece looks snowy-white, but after shearing the skin reveals black or red areas (no matter how small), the llama cannot be dominant white. According to Dr. Graham's experience, shearing always reveals colored spots on the skin of llamas with uniformly white fleeces.

Dr. Graham cautioned me not to automatically assume that white alpacas with pigment missing around the eyes are dominant whites. Since many do have dark eyes and produce, so I've been told, white crias in the first generation, the existence of the **W** locus for alpacas should nevertheless still be considered a possibility. In "Some Educated Guesses on Color Genetics of Alpacas," Dr. Sponenberg wrote: "Most white alpacas produce white or pale offspring following mating to any color, indicating that dominant mechanisms for whiteness are relatively common among alpacas" (*The Alpaca Registry Journal*, Volume VI, Number 1).

Let's go back to our colors. In an early chapter, we briefly discussed various shades such as fawn or blue. They are created by separate switches (genes) affecting our two basic pigments. We call these *dilution genes*. The animal has pigment cells — they don't, however, express full color due to the various dilution mechanisms. A combination of these genes can dilute hair or fiber to the extent of making the animal appear white.

What about patterns, such as a white Greyhound with reddish-brown spots described as "markings" by breeders? You probably (and incorrectly) think of this dog as white with brown markings. His base color is red, even if the red spot is the size of a dime and is the only one on his otherwise lily-white body. To be precise, he is red with one giant white "mark" or spot.

A BREEDER'S GUIDE TO GENETICS
Relax, It's Not Rocket Science

This Borzoi presents an excellent example of a red dog with the canine piebald pattern. The remaining red areas are well-defined, irregular, and randomly placed. The amount of white can vary considerably in size from one dog to the next
(photo by Barbara Ewing)

In the case of white patterns, the white areas are genetically directed not to produce pigment cells. *Modifier genes* then influence the extent of these patterns, which can vary from a tiny area to large patches (piebald). The entire coat may be white due to a spotting gene coding for extreme white piebaldism. The true color of a completely white Greyhound, for example, will forever remain a mystery to you unless you know the colors of the preceding generations and/or the offspring. The number of genes coding for white patterns and spots as well as their nomenclature varies from species to species.

This presents the inheritance of color in a very, very tiny nutshell.

I know you will be fascinated to find out about the functional connection that exists between pigment cells and the central nervous system. Dr. Nina R. Beyer shared information from a very informative book with me (I referred to it in previous chapters) concerning coat patterns. Written by a group of authors, a chapter in *Genetics and the Behavior of Domestic Animals* (1998, edited by Temple Grandin of Colorado State University) informs the reader about an experiment conducted in Russia on the wild red fox. After twenty years of selective breeding, "the experiment succeeded in turning wild foxes into tame, border collie-like fox-dogs." These foxes were found to have changes in their hormone profiles, shedding patterns, and "developed black and white patterned fur." The authors quote Popova, et al, (1975): "The tame foxes had higher levels of the neurotransmitter serotonin." Serotonin is known to suppress aggression and feelings of depression.

There is speculation that the depigmented coats of some guard dogs bred to protect livestock correlate to higher serotonin levels, accompanied by little or no predatory instinct.

Many variables are at work here, and drawing conclusions on dog behavior based solely on the fox study is not a wise idea.

Grandin and Deesing point out: "It is interesting that the two calmest breeds of cattle, the Hereford and the Holstein, have completely depigmented white areas on their heads." The authors also report that <u>excessive</u> removal of pigment has been implicated in animals being "more difficult to handle and more nervous."

Slightly off topic, but pertinent to the issue of temperament, is information that Dr. Graham shared with me: brain volume or weight is reduced as animals are domesticated. It is believed that brain cells associated with the "fight or flight" mechanism are lost. Furthermore, large brained, extremely intelligent animals would be more difficult to handle in commercialized operations. Very intelligent pets need challenging "work," or they become troublemakers out of sheer boredom. I always find it somewhat disconcerting that so many breeders (and authors) confuse obedience with intelligence. A highly intelligent animal may, of course, be very obedient — but a very obedient animal may not necessarily be intelligent.

In the next chapter, Dr. Beyer's third paragraph in her discussion of the Merle locus puts genetic mechanisms coding for color in the proper perspective. Before we can pursue that thought, we must discuss another issue.

Although I'm reluctant to take a "side tour" just as I begin my explanation of the color loci, you should learn one more concept to truly understand color inheritance.

You have learned and mastered the concept of how one allele <u>in a single pair</u> can be dominant over the other or recessive to the dominant one, such as Bb. *Epistasis* also involves dominance of one allele over another, only this time it does not involve its allelic partner at the same locus. Let's use two loci — **B** and **D** — to explain this concept. You will learn later that a **BB** or **Bb** dog can only be black if it is **DD** or **Dd** as well. The effect of **BB** is masked by **dd**, even though **dd** does not occupy the same locus as **BB**. A **B- dd** dog is blue. A gene or combination that changes the expression of another gene <u>at a different locus</u> is called *epistatic*. The gene or combination that is being changed (diluted or masked, for example) by a gene or combination <u>from another locus</u> is called *hypostatic*.

The tricky part is that the **d** allele is recessive to **D** — but carried in the homozygous form, **dd** is nevertheless *epistatic* to the dominant **B** at the **B** locus.

When you hear or read the word *epistasis*, just remember that more than one locus is involved.

To use our wrestlers again: the heavyweight (**B**) will always pin the flyweight (**b**) from his own gym, but <u>two</u> flyweights (**dd**) from the <u>**D** locus gym</u> will take the starch out of **B** in a hurry by ganging up on him together.

You still don't quite understand this? I didn't either when I initially encountered the subject. Not everyone comprehends the information on color genetics the first time. Don't give up! Read the next chapters several times. Then study the more detailed information in *Genetics of the Dog* (Willis) and *Equine Color Genetics* (Sponenberg). Trust me — you will eventually understand the concepts!

I highly recommend that novice breeders of all species interested in color inheritance read Dr. Sponenberg's book. Not all of the genetic mechanisms described there will apply to other mammals in identical ways. However, his book on horse and donkey colors clarifies the basic mechanisms of color inheritance, while at the same time addressing more complex concepts than the format of this handbook permits. I consider it an absolute necessity to be included in the libraries of serious horse and donkey breeders.

Don't let my choice of the word *complex* frighten you. Dr. Sponenberg's book is very readable. Its contents triggered the "lightbulb moment"

I mentioned in a previous chapter. The author clarifies the <u>interaction and orderly progression</u> of the various loci towards a final phenotype.

None of our Earth's multitude of species evolved in a genetic vacuum. I know, I know — I've made this point several times already. It just can't, in my opinion, be emphasized enough! Understanding the inheritance of dog colors will also help you comprehend inheritance mechanisms in llamas, sheep, horses and other species. Some of the larger mammals enjoy colors and patterns inherited in an identical fashion to those of fancy mice. Fancy that!

Entire books have been devoted to the subject of color inheritance for just one species. It is totally beyond the scope of our work to go into such minute details. Instead, I will present a rough outline of the various loci as presented in *Genetics of the Dog*. To truly understand the subject in relation to your own animals, you should explore research specific to their species.

Our knowledge of camelid color inheritance is still incomplete. Nevertheless, I'll present possible llama – and alpaca loci and genetic mechanisms as they have been proposed by several knowledgeable people in the field.

Okay, you had your fun. We flitted around and touched on all kinds of interesting little tidbits. Now comes the nitty-gritty stuff.

Chapter Thirty

COLOR LOCI

CHAPTER THIRTY

Color Loci

"...color genetics is inherently complicated. Unless viewed in its details, it is not going to provide many answers that will be of practical help to breeders."
D. Phillip Sponenberg, D.V.M., Ph.D., *The Alpaca Registry Journal*, 2001

Remember to think of each gene coding for color as residing at its own *locus*. A locus can be described as a genetic "address." Also remember that the individual animal can only have two genes (alleles) at each locus — one inherited from the sire, and the other from the dam. If you don't recall the details of multiple allelism, refresh your memory by reviewing the information about mitten color choices in Chapter Three. When you read the word *epistasis*, remember that more than one locus is involved.

Once again, authors differ in their selection of symbols. This is perfectly acceptable, and does not mean one is correct and the others are wrong. Let me give you an example using the **C** locus. You will also see this described as the *C series* or the *albino series*. In most species it doesn't even carry a true albino allele, but is called by that name, nevertheless.

Clarence C. Little, Sc.D., uses the symbol $c^e c^e$ for extreme dilution. Dr. Julie L. Parver Koenig wrote it up as **CC** in a packet handed out at a workshop (*Genetics — From Buying to Breeding*, 1990). She lists "no dilution" as **C+C+**. Dr. Sponenberg describes the combined alleles creating cremellos as $C^{Cr}C^{Cr}$. As I've explained in "Alphabet Soup," authors always define their own nomenclature, so don't be put off by that practice.

Please remember that information about your chosen species or breed may be slightly different or much more complex than what is presented here. This is just the basic blueprint. Think of it this way: once an architect learns how to read blueprints, he will understand the prints for a ranch dwelling as well for a Colonial mansion or an industrial warehouse. Grasping the basic information in this guide will give you intellectual access to a mountain of research and observations made by scientists and experienced breeders.

If you are not a dog breeder, please don't ignore the information on dog color inheritance. Yes, color loci may vary somewhat from species to species. Yes, a pattern gene may be dominant in one species and recessive in another. However, if you learn to read the color "blueprints," you will be able to skip from dogs to mice to alpacas without missing a beat.

The Agouti Locus in Dogs

One of the most important loci — and usually the first one mentioned by all geneticists — is the **A** or **Agouti** locus. Depending on which author you read, there are 4 to 5 alleles. Remember that each individual can only carry two alleles at this or any other locus. Specific alleles can be and are indeed lost in various breeds (genetic drift).

Malcolm B. Willis tells us that in the case of *multiple allelism* the alleles are listed in order of dominance. For example, if an author lists **A**, a^y, a^g, a^s, and a^t in that order, it means that **A** is dominant over all the others, while a^y is recessive to **A** but dominant over a^g, a^s, a^t and so on.

Think of the alleles as NFL football players, drafted in the order of playing ability by various teams. The best player is naturally drafted first. The second draft choice is not as good as the first, but is better than all the ones following him, and on down the line. This analogy isn't great, since in many breeds one color is not "better" than the other, but you get the picture.

Think of the Agouti locus as the "boss" or control center of the entire color scheme. The other loci can be loosely defined as supporting cast members. The Agouti locus initially determines how and to what extent dark color is distributed on an animal genetically programmed to produce and express black or red. I've already mentioned how other loci can phenotypically change those choices (epistasis).

Please note that all five alleles are clustered under the term *Agouti locus*, although only one of the five alleles produces the typical "banded" agouti pattern. As a novice student of genetics, I found this to be a most confusing state of affairs. Since scientists rarely take the diminished capacities of the scientifically challenged into consideration, we should accustom ourselves to this *agouti allele at the Agouti locus* business.

The *Agouti* alleles as listed by Willis:
  A = Dominant black
  a^y = Dominant yellow (golden sable)
  a^g = Agouti (wolf grey)
  a^s = Black or liver saddle markings
  a^t = Bicolor (saddle markings extended over most of the body)
  a^s and a^t also carry tan markings on head and legs
  a^g is also described as wild-color (some geneticists use the term *original type*)

Although a^s mentions liver, it takes another locus to modify the original black to liver. Likewise, the black pigment formed by <u>any</u> of the other alleles can be modified in such a way.

While **A** allows distribution of dark pigment over the whole body, all other alleles restrict it to various degrees. The a^y allele produces variations of color ranging from red to yellow to tan, with dark pigmented hairs scattered in the coat. The extent of the dark pigmented hairs can vary considerably. In some animals, they may be hardly noticeable. Researchers also believe in the *dual expression* of the combined a^y and a^t alleles, resulting in distinctly different phenotypes for $a^y a^y$ and $a^y a^t$ animals in some breeds.

Study Sue Ann Bowling's web site at **http://bowlingsite.mcf.com/Genetics/GolorGen.html**. Bowling points out that breeders of Shetland Sheepdogs and German Shepherd Dogs established a recessive black in their breeds. Since we don't find dominant **A** at the Agouti locus in other species, Bowling speculates (along with other breeders) that **A** may exist at a separate locus. A scientist told me that he considers the existence of a dominant black allele at the Agouti locus to be "biochemically impossible." Indeed, when you look at the progression from light to dark ($a^y \rightarrow a^t$), it doesn't seem logical to place black as the dominant allele at this particular locus. However, since Dr. Willis's information (originally proposed by Dr. Little) works for most breeds, I will continue to refer to **A** as part of the Agouti locus in this book.

Extension Locus in Dogs

Dr. Willis does not list the *Extension* locus right after the Agouti locus. I did because of the close interaction that takes place between these two loci. He quotes the research and allelic order first established by Dr. Little.

The alleles listed are:
- E^m = Superextension with dark mask
- E = Extension without black mask
- e^{br} = Brindle or partial extension
- e = Restriction

The term *Extension* refers to the extent that dark pigment is allowed to be expressed by each allele.

E^m codes for full expression of the amount of black in the coat <u>as allowed by other genes</u>. For example, if a dog is **AA** or possibly **A** in combination with other alleles recessive to it (such as Aa^y), the E^m allele supports the Agouti alleles in their quest to express themselves as black. According to Dr. Willis, $a^y\ a^y$ dogs are red-yellow or sable with a black mask when E^m is present.

All subsequent alleles at the Extension locus either allow only partial expression of black — such as **E** permitting the extension of black over the body except the mask — or none at all. The homozygous form of the **e** allele (**ee**) causes all black pigment to disappear, except on the nose and other pigmented skin. Geneticists would say that **ee** is *epistatic* to the Agouti alleles, while the Agouti alleles are *hypostatic* to **ee**. An **AA ee** dog is therefore phenotypically red (yellow) rather than black. Yellow Labrador Retrievers are examples of the **AA ee** combination.

Some breeders deny the existence of the e^{br} allele as part of the Extension locus and feel that results in their particular breed support their theory. If you want to select for or against brindling, I strongly suggest you discuss this issue with a knowledgeable person in your chosen breed.

Again, check out Bowling's web site. She supports the theory that e^{br} does not belong to the Extension locus and bolsters her argument with concrete data and examples.

The B and D Loci in Dogs

The **B** locus is easy to understand — it consists of only two alleles. Dominant **B** allows black pigment to be formed; the recessive **b** modifies black to chocolate brown or liver (the color, not the delicious dish smothered in sauteed onions). In its homozygous form (**bb**), it affects coat pigment and eye color. Review: **BB** and **Bb** = black, **bb** = chocolate brown or liver. If you recall, chocolate was once considered a separate pigment in mammals. It is not.

Remember, **B** alone will not produce black pigment over the entire coat. The Agouti and Extension alleles determine the extent. The $a^s a^s$ **EE BB** dog is a red dog with black saddle markings. Again, an **AA ee BB** Labrador Retriever is yellow, not black. You can easily figure out the genotypes and phenotypes of various "paper breedings" by trying out different combinations using alleles from the loci we've discussed so far. Scientists have speculated about so-called *rufus polygenes*. These genes, according to Dr. Willis, either lighten or darken the coats of brown/liver/red dogs.

Here's another piece of the puzzle: the **D** locus. Dominant **D** causes no change in dark colors; recessive **d** dilutes dark colors. **DD** and **Dd** have no effect on other loci; **dd** washes out the color already present. Unlike **ee**, **dd** also lightens the color of pigmented skin as well as eye color. The description of the **B** and **D** loci presents another opportunity to review epistasis. Recall that **BB** can only be phenotypically expressed as black if one **D** allele is present at the **D** locus; therefore it is hypostatic to the **D** locus. If **D** is not present, the **A- E^m- BB dd** dog will be blue. Once again, the Agouti and Extension loci must carry the appropriate alleles for black before **B** and **D** even enter the picture.

Owners of Whippets and some other dog breeds can identify this locus in their animals. Their black Whippets carry either **DD** or **Dd** at the **D** locus. The black **dd** Whippets will look blue. Red will be diluted to fawn by **dd**.

Greyhound and Whippet breeders (as well as others) speak of *blue brindle*. These animals express the normal brindling patterns. The black "stripes," however, have been diluted to blue by a double dose of dilution alleles at the **D** locus. Eye color, as we already discussed, has been lightened as well. It's absurd for a parent breed club to permit all coat colors in its breed standard, yet disqualify animals with very light eyes.

This section reminds me of a fellow breeder's young daughter, who got in trouble with her teacher when she argued that dogs can be blue — which was the answer she had marked as correct on a multiple choice reading comprehension test. I bet that teacher never heard of lilac-colored mice, either.

The Grey, Merle, and Ticking Loci in Dogs

The addition of the Grey, Merle, and Ticking loci adds spice and variety. **GG, Gg,** and **gg** <u>will all be born black</u>, providing the other loci code for it, of course. While the first two genotypes gradually transform the black to a bluish grey, **gg** remains black. Note that the allele coding for grey is dominant! It is easy to confuse the effects of **dd** and **G**. A blue dog might look grey to the uninitiated observer. To describe such a dog as grey is genetically incorrect.

The Kerry Blue Terrier serves as an excellent example. Its name is somewhat misleading, as it does not derive its "blue" color from the **d** allele of the **D** locus but rather from the **GG** and **Gg** combinations of the greying series. Its amazingly homozygous genetic color blueprint reads **AA BB CC DD EE mm SS tt**. Only the **G** locus offers choices of **GG, Gg,** and **gg**. All Kerries are born black; the **gg** ones will remain so for the rest of their lives.

The important concept to understand is that there is no "recessive black" at these loci. The choices at the **G** locus are **grey** or **not grey**. Likewise, the choices at the **D** locus are **blue** or **not blue**. In other words, the choice is <u>never</u> **black** or **grey** at one locus. This is probably one of the most important concepts to understand.

The homozygous combination of the dominant merling gene (**MM**) produces defects. Breeders whose chosen breed carries the dominant allele of this gene are well advised to investigate the details of possible genetic repercussions. Since it is a dominant mechanism, producing an **MM** animal should be easy to avoid. I thought we'd give Dr. Willis a rest (figuratively speaking), so I asked Dr. Nina R. Beyer to explain the details of this rather complex locus.

From personal correspondence from Nina R. Beyer, V.M.D., September, 2000:

The merle gene is a dilution gene; its effect is to dilute or lighten the color of the coat. It works in patches of coat, leaving adjoining patches the "original" color. It doesn't work on the tan-colored points on the face, legs, etc. so these also retain their color. If we call the merle gene **M** (capitalized because it's a dominant gene), a fully pigmented dog would be **mm** and a merle dog would be **Mm**. Knowing this, you can quickly see that breeding two non-merle dogs will never produce a merle puppy (**mm** x **mm** > only **mm**), and that breeding a merle to a full-colored dog will produce, on average over a large number of breedings, half merles and half full-colored dogs (**mm** x **Mm** > ½ **Mm** and ½ **mm**). What happens if you breed two merles together? You'll get about one-fourth full-colored dogs, one-half merle dogs, and one-fourth dogs with the genotype **MM** (**Mm** x **Mm** > ¼ **MM**, ½ **Mm**, and ¼ **mm**). And what do **MM** dogs look like? These so-called double merles are white, usually deaf, often blind, are occasionally born without eyes at all, and usually have other physical problems as well. This is why some breed clubs prohibit breeding merle to merles; they don't want to knowingly chance producing 25% dead or defective puppies. (Denise has seen the sad results of such merle-to-merle breedings in collies, and strongly agrees.)

In Australian shepherds, a liver/red dog displays the red-brown variant of the black gene; if it's carrying the merle gene, it's called a red merle. A black dog with the merle gene is called a blue merle because the grey patches among the black patches really do look blue. In dachshunds, the markings called dapple are the very same merle gene. In Shetland sheepdogs, there is no acceptable liver color. Both the tricolor sheltie (black with tan points, usually Irish marked with white) and the bi-black sheltie (black with no tan) carrying the merle gene will be blue merles; of course, the second one is called bi-blue. If a sable sheltie is a merle, the patches won't be very obvious. If it wasn't for the fact that many (but not all) sable merles have particolor blue eyes, they could easily be mistaken for regular sables. If two sable merles are bred, one-fourth of the resulting puppies could potentially be double merles with the same problems listed above. To avoid producing sable merles in the first place and prevent potential heartache in the following generation, most sheltie breeders insist on only breeding merles to tricolors (or bi-blacks).

Genes don't do just one thing. The genes controlling the presence of pigment don't exist just so we can have colors; melanin is involved in the neural crest cells that form the hair cells in the ears, and its absence damages hearing, just like the retina's function is damaged by lack of pigment. That is why something that seems simple, like color, really isn't.

Dog breeders refer to small colored spots as *ticking*. These "freckles" are inherited in a dominant fashion — they are sometimes not visible due to a dark base color, or a lighter colored long coat. Pointers, setters, spaniels, Dalmatians, and some of the hound breeds furnish good examples of dogs expressing the ticking factor. Ticking, at least in the Borzois and Whippets I know, seems to become more pronounced as the dogs age (Denise is nodding her head "yes"). A wet coat will reveal a ticking pattern rather nicely (and at times surprisingly) on a heavily coated dog — the spotting shows up clearly on the skin. It's amazing what you see on a "white" dog when it gets a bath!

Spotting Locus in Dogs

White patterns are the most interesting facets of color inheritance. Some species, as well as breeds within a species, do not carry the appropriate alleles for patterning to occur. For example, if a pattern is passed on as an autosomal recessive, and every single animal in a certain breed is homozygously dominant at that locus, it just isn't there to surface. Natural and human selection pressure has seen to it that brown bears and black Labrador Retrievers are all **SS** (self-colored). Other species and breeds carry more than one pattern gene. The important thing to remember is that a single dog can be the recipient of two distinctly separate alleles coding for patterns at the **S** locus. Just looking at the phenotype of such an animal will not always reveal the full genotype.

Let's review: contrary to popular opinion, spotted animals are not white with colored spots. They are colored with certain areas stripped of their pigment. Sometimes the white "spot" extends over the entire body (extreme white spotting).

Dog geneticists commonly assign the allele coding for such a huge "spot" to the **S** or *Spotting* locus. Recessive alleles at the **C** locus (Albino series) may also produce a completely white coat. However, extreme white piebalds must be viewed as totally separate entities from **C** locus whites. Dr. Willis and others who have studied the subject in depth agree on four alleles directing the genetic traffic at the **S** locus.

These are presented in *Genetics of the Dog*:

In order of dominance:
S = Self colour or totally pigmented surface
s^i = Irish spotting involving a few definite areas of white
s^p = Piebald spotting
s^w = Extreme-white piebald

(The spelling of the word "color" as "colour" identifies the author as British.)

Irish spotting, named after the identical patterns found in rats, refers to a very distinct pattern involving white markings on several body parts, including the muzzle, feet, chest, and end of the tail.

The piebald allele extends irregular white markings or patches over large parts of the body. Piebald spotting can vary quite a bit in its extent. Dr. Willis points to the beagle as an example. Remember that modifiers play a large role in the expression of this particular allele.

Extreme-white piebald refers to completely white dogs. Again, these are not to be confused with **WW** or **Ww** animals in other species, or white animals homozygous for the recessive alleles at the **C** locus. (Geneticists studying color inheritance in dogs do not believe that **W** exists in that species.) Unlike **W**, the extreme white piebald allele (s^w) is recessive to self-color (**S**), hence the shock some breeders experience when self-colored parents produce completely white offspring.

Although s^w is recessive to **S**, the homozygous ($s^w s^w$) form is epistatic to other loci. Such a double dose would make the otherwise black **AA EE BB** dog white.

The C Locus in Dogs

Very few species produce albino as a regular genetic pattern (wild albinos experience survival difficulties and usually don't grow up to reproduce). True albinos, such as Angora rabbits, have red eyes. Nevertheless, Dr. Willis, as well as other scientists, refer to the **C** locus as the Albino locus or series.

Dr. Willis lists the following alleles:

C = Colour factor which allows melanin to be formed
c^{ch} = Chinchilla
c^d = White coat with black nose and dark eyes
c^b = Cornaz coat with blue eyes
c = Albinism with pink eyes and nose

There have been disagreements among geneticists about the existence of c^d, and the c^b allele is considered rare.

The alleles are arranged in order of "severity" of dilution. **CC** codes for full expression of color without any dilution whatsoever. The recessive **C** locus alleles generally affect red pigment more than black, <u>thus making it possible for black animals to suppress expression of the dilution alleles in their phenotype to various degrees</u>. Black pigment may be diluted to a silvery color (horse breeders call it *smoky*). Red pigment is diluted to cream by the Chinchilla allele. Some authors, disregarding the occurrence of true albinism altogether, assign white coat/blue eyes to the **c** allele. The impact of the various alleles at this locus depends very much on which diluting genes are already present at other loci. Some breeds exhibit very pale cream or almost white coats as the result of the **ee** $c^{ch} c^{ch}$ combination.

Just as in horses, the **C** locus alleles may be incompletely dominant in some breeds, with **C** c^{ch} producing moderate dilution.

Regardless of chosen nomenclature or disputes over the exact cause of certain phenotypes, all authors agree that the downward progression at the **C** locus proceeds from full pigmentation to its total absence.

While discussing genes coding for loss of pigment, I should mention the Roan locus. Various authors can't agree that it exists in dogs. This is, in any case, discussed later in the chapter.

Horses, Cats, Sheep, Rabbits, Mice, and Goats

I want to touch briefly on interesting aspects of color genetics in several other species. Horse owners should definitely study the initially overwhelming number of mechanisms that control the numerous shades of color in that species. The effects of the loci producing them have been extensively studied and documented — a wealth of research material is available for the lucky horse breeder.

The color and pattern diversity of domestic cats must eventually intrigue any cat breeder enough to investigate the genes behind their inheritance. I've provided information on a very readable book I found on cat genetics in the reference section.

Denise, a self-described "information junkie," found a web site for cat breeders. Orca Starbuck and David Thomas give a detailed description of cat color genetics at their web site, **http://www.fanciers.com/other-faqs/color-genetics.html**, as well as their e-mail addresses if you have further questions.

Sheep breeders will also find an abundance of information available on color genetics. Beginners might want to start by reading Susan Mongold's article "Color Genetics in Icelandic Sheep" in *The Shepherd* (June 1997, Vol. 42, No. 6, pp 11-16), or access her website at **http://www.icelandicsheep.com/genetics**. Mongold's well-presented color/pattern table shows 21 possible combinations that Icelandic sheep can inherit. These hardy ruminants ob-viously carry dominant white and several other interesting patterns, including one I would call "original type" — breeders of Icelandics named it *Mouflon*. Color is determined by the two choices at the **B** locus — black or brown. Apparently, there are no red Icelandic sheep.

Jacob sheep, an ancient breed, express an interesting black and white spotting pattern. Occasionally, the black areas of the fleece are brown or "lilac." Handspinners love the spotted Jacob fleeces.

Mouse breeders also developed a lilac specimen in their mouseries. I learned from breeder W. MacKintosh Kerr that the lilac-colored rodent "is a Chocolate mouse carrying a double dose of the blue factor (**dd**)" (*Color Inheritance in Fancy Mice*). A chocolate mouse is, as we know by now, a black mouse carrying **bb** at the **B** locus. Interestingly, the author mentions that the lilac mouse has dark eyes. The definition of "dark" is, of course, relative.

The author of *Cat Genetics* (1977) tells readers: "If one wants to make a lilac, there is a lot to be said for mating the blues to the chocolates..." In order for this to work, blues must carry chocolate (**aa Bb dd**) and chocolates must carry blue (**aa bb** Dd). The lilac babies — cats or mice — will be **aa bb dd**. Enemies for millennia — united by sharing their genetic mechanisms!

As I mentioned in an earlier chapter, black is recessive to red in cats and many other species.

After reading these last two paragraphs, my husband expressed the fervent wish that no "normal" person will ever read this book. I told him not to worry. "Normal" people do not have the desire to breed anything and will not spend money to purchase this guide!

Rabbit color genes are well known and documented for interested breeders. Apparently, a computer program exists that provides possible color/ pattern results for proposed matings. I understand that goat color inheritance is extremely complex (is a high IQ a prerequisite for breeding goats?).

Camelid Color Genes

The array of camelid colors is positively dazzling. From white to black, from self-colored to pintos to appaloosas — the combinations of colors and patterns seem endless. Continued research is needed to determine color inheritance in llamas and alpacas. Several people have provided valuable observations, but much remains speculative at this time. The genetic mechanisms coding for the many fascinating variations will not be easy to determine.

Several aspects complicate the study of camelid color inheritance. Let's initially focus on alpacas. To register an alpaca, a breeder must prove

A BREEDER'S GUIDE TO GENETICS
Relax, It's Not Rocket Science

parentage through DNA testing. The Alpaca Registry Inc. provides breeders with an attractive color chart of 16 "basic representative colors in alpaca fiber" (see Table A). The registry freely acknowledges that there are "many variations of these colors which can amount to more than 22 natural shades produced in the alpaca fleece." Breeders are encouraged to choose a designation most closely matching the actual fiber.

TABLE A — Key to ARI Natural Fiber Colors	
White	W
Beige	B
Light Fawn	LF
Medium Fawn	MF
Dark Fawn	DF
Light Brown	LB
Medium Brown	MB
Dark Brown	DB
Bay Black	BB
True Black	TB
Light Silver Grey	LSG
Medium Silver Grey	MSG
Dark Silver Grey	DSG
Light Rose Grey	LRG
Medium Rose Grey	MRG
Dark Rose Grey	DRG

Fleece colors can undergo subtle changes due to environmental or genetic influences. One breeder confessed to me that a chocolate brown female cria "turned" black, and was indeed jet black when she was sheared for the first time. Several years later, her fiber remains a deep black — and she's registered as a dark brown!

We have thus identified the first difficulty when trying to determine modes of inheritance by studying data from the registry: owners often register very young crias whose fleeces have not necessarily revealed their "true" and final colors.

Sun exposure and possibly nutrition can impact an animal's fiber phenotype. Horses, for example, darken when they are well fed. The sun can bleach the tips of llama – or alpaca fleece. Although the registry chart urges breeders to "take the clip of fiber as close to the skin as possible," there is no guarantee that all owners submitting registration papers actually do that.

Registration certificates do not always list the crucially important identification of "point" colors. A red (brown) alpaca with black ear-tips, tail, and lower legs is, genetically speaking, an entirely different creature than a solid red alpaca. A self-colored medium fawn shares some — but not all — genes with a fawn expressing the original type pattern (white or cream belly and anal region). A geneticist needs to know such details to establish how specific genes are inherited.

The registry can only identify a white animal as white. The geneticist must probe further. Is it a dominant white or a dilute white or possibly an extreme white piebald? Minimal expression of genetic mechanisms (and not identified as such on registration certificates) further confuses the issue — so do authors of articles or books who lump all white llamas or alpacas together as belonging to a genetically uniform group. You will never sort out patterns of inheritance that way.

"Shades" of white are a separate issue, and the two are not to be confused. In "Some Educated Guesses on Color Genetics of Alpacas" (*ARI Journal*, Spring 2001), Dr. Sponenberg mentions the various "shades" of white in the alpaca color charts and advises "lumping" them together to "sort out the big picture of color genetics."

The subtle variations between individual "white" camelids are different than the concepts of *dominant white* or *extreme white piebald*. In other words, two extreme white piebald animals may be two different "shades" of white.

These variations may not necessarily be of genetic origin. Soil content and choice of bedding can give a snowy white fleece a beige, rosy, or greyish hue. In our study of genetics, decorator terms such as "linen white" or "off white" can be ignored from a practical standpoint.

One breeder recently expressed frustration over the fact that a rose-grey cria sired by her stud was registered as brown by the dam's owner. "You had to part the fleece to see the true color pattern," she insisted.

Well, that's the problem in a nutshell. The study and recording of data as observed <u>on the animals</u> will give camelid breeders a more complete genetic picture.

Is the registry information, as one breeder suggested to me, totally worthless as a tool to provide valid and useful information? Not at all. Registry data, for example, tells us that a white cria out of <u>two</u> colored parents is positively not a dominant white. A trained geneticist would certainly be able to suggest meaningful changes as far as color identification and registration are concerned. However, we must recognize the fact that any system is only as good as the input provided by owners and breeders.

Llama breeders have the added complication of unverified parentage in their stock. Carelessness, human error, and possibly (sorry to say) just plain fraud may invalidate some of the llama data from a strictly scientific standpoint. While there probably are differences, I imagine that the genes coding for llama and alpaca colors and patterns are very similar. It certainly behooves llama breeders to pursue research on alpaca colors and vice versa. Please recall Dr. Wheeler's research using alpaca and llama DNA!

When I speak of camelids, I am referring to llamas and alpacas in the context of these particular chapters. I simply choose this term for convenience.

<u>Please remember that the personal opinions expressed in this chapter are based on speculation and my own observations. They should not</u>

be read, interpreted, repeated, or quoted as a definitive study of camelid color genes.

I did not have the pleasure of speaking with Dr. Koenig. Dr. Graham made it clear (in personal correspondence) that she is always receptive to new discoveries and insights. The title of Dr. Sponenberg's article "Some Educated Guesses on Color Genetics of Alpacas" (*ARI Journal*) gives us a hint that, unlike his research in horses, dogs, sheep, and goats, the study of camelid color genetics is still in its infancy. Sponenberg states that comparatively little camelid data is available, although he makes it clear that "the hard evidence" he has seen supports his reported findings.

Readers need to understand that determining the inheritance of camelid colors will be an ongoing process for many years to come. Experts will revise and add to their information as the scope of their research broadens.

To the readers who ignored my advice and skipped over the information on dog color loci: flip back to those pages at the beginning of the chapter. Let's have none of this sneaky stuff!

The Agouti and Extension Loci in Camelids

You will understand what I'm saying about alpacas and llamas much easier once you've studied the well-established genetic mechanisms controlling color genes in, for example, dogs or horses. Readers who enjoy very limited knowledge of color genetics are prone to misinterpret basic information. One confused author, who shall remain nameless, claimed Dr. Koenig assigned self-black dominant status in camelids. She did no such thing. She lists original type as the most dominant of the Agouti alleles, followed by bay (red-brown with black legs and head) and seal-brown bay (born black, lightens to dark brown). She clearly lists black as the most recessive allele at the Agouti locus, with the **B** locus determining its expression as either black (**B**) or brown (**b**) — brown being the chocolate-brown of mouse color genetics fame (*Genetics — From Buying to Breeding*).

The author of the previously mentioned article obviously only saw that **B** (black) is listed as dominant over **b** (brown). He did not understand the difference between **Agouti** locus brown (really red) and **B** locus brown, which is actually ...

Well, those of you who have done their homework and have read the entire chapter up to now will know how to finish that sentence (and will not misquote Dr. Koenig). Both Dr. Koenig and Dr. Sponenberg believe red as well as red with black trim (with rare exceptions) to be dominant over black in llamas and alpacas.

Sponenberg concluded from his research that "the Agouti locus appears to be a major determinant of color variation in alpacas."

Please notice that Sponenberg uses capital **A** to label all Agouti alleles. This does not negate the order of dominance. In other words, A^a is just a different symbol for **a**. For example, the homozygously recessive combination **aa** is written as $A^a A^a$ under such a system.

I personally think this alphabet game is very confusing for beginners, but you'll get used to it as time goes on. For the sake of consistency, I will continue to use these symbols while referring to camelid color genes — except in my initial quotes from Dr. Koenig's work. It helps if you focus your eyes on the superscript. Let me explain: when you read, for example, $A^a A^a Rn^R Rn^+$, focus on the $aa\ R+$. This is a black animal with a roan pattern.

The following Agouti alleles present the most recent update (2003) in Dr. Sponenberg's research on alpaca color genetics:

Agouti Locus
A^T - tan
A^+ - tan with minor trim
A^r - red with black trim
A^b - bay
A^t - black with light belly
A^m - mahogany
A^a - black

Sponenberg writes: "It is important to note that each animal can have only two of these variants, and generally only express the paler of the two." If you read the explanation on the order of dominance at the canine Agouti locus (multiple allelism), you are already familiar with the concept.

Dr. Sponenberg's proposed order of alleles at the Extension locus is as follows:

Extension Locus
E^D - dominant black
E^+ - wild type (neutral), as determined by Agouti
E^e - recessive red or tan

As is true for other species, the alleles at the Extension locus are epistatic to the Agouti alleles.

Some alpaca breeders assume that an inky, deep (non-fading) black is the result of a dominant gene, while a recessive gene codes for a fading, washed-out-looking black fleece.

Let's be careful not to jump to hasty conclusions regarding the genetic control of fading versus non-fading blacks.

Dr. Sponenberg points out that in other species, "dominant black is somewhat 'weaker' than the Agouti recessive black, and is frequently an off-black or very dark brown instead of true black."

In *Equine Color Genetics*, Sponenberg devotes an entire section to the definition, classification, and genetic control of *shade*. He refers to the subject in an alpaca workshop hand-out: "In other species, the fading or off-black types include the dominant blacks (so if you have been thinking that these are 'jet black' then think again), or possibly they are recessive black carrying one dose of the recessive allele at the Brown locus." In other words, the intensity of black pigment is not necessarily linked in any way to a dominant or recessive mechanism for black at the Agouti or Extension locus.

Dr. Sponenberg states that dominant black is extremely rare in alpacas, "if it exists at all." From a practical standpoint, it may be advisable to presently ignore this allele altogether.

Dr. Graham speculates about a dominant black found in llamas. When I asked her about it, Graham clarified the matter further by differentiating between "recessive black" (tends to sunburn to red on the top) and "shiny" or "intense black" with reddish areas appearing in the fleece. She believes these *calicos*, as she calls them, to be under dominant genetic control.

Dr. Graham made it clear (in personal correspondence) that a calico is not the same genetically as other blacks. It "shows red in its flanks" which is not the same as a sunburned fleece. She further clarified that her belief in a dominant mode of inheritance is based on anecdotal evidence only, and is certainly not "written in stone." Graham does not believe that the calico gene in llamas is sex-linked as it is in cats.

In reference to the E^+ allele, the term *wild type*, also called *original type*, has nothing to do with the animals still being "wild" (not domesticated). It refers to the original color/pattern expressed in that species or breed. In dogs, for example, the wild type color is the wolf-grey agouti pattern still found in a number of breeds. In the case of alpacas, it would be the color of the vicuña. This allele is neutral. The vicuña pattern is only expressed if the corresponding Agouti and Pangaré alleles also code for it. Such alpacas are $A^+\text{-}E^+\text{-}P^{PA}\text{-}$.

Expression of the E^eE^e combination is similar to what we find in horses: "At the Extension locus, the recessive E^e allele results in uniformly red horses without black points" (*Equine Color Genetics*). The Extension locus reds are chestnut horses. The ee combination (written up as $E^e\ E^e$ by Dr. Sponenberg), as we learned from Dr. Willis, suppresses all black pigment in dogs. Likewise, red alpacas without black trim could possibly be $A^r\text{-}E^eE^e$. A beige-yellowish-red animal without any black may be $A^aA^a\ E^eE^e$. This alpaca would actually be genetically black (similar to yellow Labrador Retrievers) and go on to reproduce as such. Bred to an $A^rA^a\ E^+E^+$ (red with black trim) partner, such a mating could produce an $A^aA^a\ E^+E^e$ (black) cria.

Let us emphasize again that the description of a red alpaca with black trim as simply red (brown) is, genetically speaking, not accurate.

The Pangaré Locus

Horses as well as other species express a genetic modification referred to as *mealy* or *pangaré*. In its most extensive form, it causes a "dramatic paleness to the ventral body color" such as "pale red or yellowish" (*Equine Color Genetics*). *Ventral*, in anatomy and zoology terms, refers to "near, on, or toward the belly or the side of the body where the belly is located" (Webster's Dictionary). This is opposite *dorsal* — "of, on, or near the back." Sorrel horses, for example, are really chestnut horses with the mealy (pangaré) effect. Black becomes seal brown.

Dr. Sponenberg identifies the pangaré gene as dominant over non-pangaré. How is his research in horses significant for camelid breeders? He finds substantial evidence that the pangaré allele exists in alpacas. Similar patterns in sheep and goats each have a separate Agouti locus allele. (So if you're stymied by camelid color genetics, think about those poor sheep- and goat breeders.) In horses and donkeys, the gene is at a separate locus and acts as an "overlay" pattern to modify the colors expressed by the Agouti locus alleles. Dr. Sponenberg believes this to be the case in alpacas ("Some Educated Guesses on Color Genetics Of Alpacas," seminar notes, 2002). A bay alpaca, for example, becomes a shaded red (brown) if it inherits the dominant pangaré allele. A tan alpaca with minor trim would express the vicuña pattern, etc. In other words, each Agouti phenotype (solid color) has a shaded variant when it is modified by the pangaré allele. If the alpaca pangaré pattern is indeed dominant, breeders must actively select for it to preserve it in the North American population.

The B Locus in Camelids

Both Dr. Sponenberg (with reservations) and Dr. Koenig list the **B** locus for llamas and alpacas — with black dominant over chocolate brown. Remember that, as in dogs, the **B** locus alone will not code for black. The black alpaca or llama would have to be $A^a A^a\ B^B B^B$ or $A^a A^a\ B^B B^b$, with $A^a A^a\ B^b B^b$ resulting in chocolate brown. Black trim is equally affected by the **B** locus. A red llama or alpaca with black trim becomes a red llama or alpaca with brown trim if it is $B^b B^b$.

Dr. Sponenberg does not totally accept the existence of the B^b allele at the **B** locus in alpacas. All the chocolate brown alpacas he personally observed had black trim (genetically impossible with $B^b B^b$).

An Australian study reported "brown" offspring from black parents (both sire and dam). No mention was made of point color or DNA verification of parentage.

A BREEDER'S GUIDE TO GENETICS
Relax, It's Not Rocket Science

Dilution Genes in Camelids

Although I refer to *diluting genes* and quote others who do, please remember that they remain a genetic mystery at present.

Dilution genes have been identified in other species. Some, as you've learned, affect mostly red pigment and allow the black areas of a coat/fleece to remain black. Others change a deep black coat to grey or blue.

At the **C** locus, Dr. Koenig differentiates, as I did in an earlier chapter, between **CC** = no dilution, **Cc** = moderate dilution, and **cc** = intense dilution. (In contrast to her workshop handout, Koenig chose these particular symbols for an article in *The Fiberfest* magazine, Vol. 3, Issue 1, Fall 1995). It does not change their meaning, nor is it really an issue. I merely mention it to spare you any confusion. I found the handout, as well as the article, very informative and well presented.

If the **C** locus applies to alpacas, animals with phaeomelanin (red) as their base pigment will have their color diluted to a much lighter tan or fawn by one recessive allele at the **C** locus, making them **Cc**. The red **cc** alpacas would have most of their color "washed out" and would be completely cream — although they often appear white to the observer. Some may be blue-eyed whites (cremellos).

As observed in other species, one dose of the recessive **c** allele may have a reduced effect — or no effect — on black pigment. As documented pedigrees expand and breeders become more knowledgeable about color inheritance, these guessing games will happen with less frequency.

A while ago, I received an interesting message from a fellow alpaca breeder who also owns horses. She shared the fact that cremello horses are not usually shown, though many horse breeders like them in their breeding sheds because they produce beautiful palominos, buckskins, and isabelos. Alpaca breeders need to read and digest this comment thoroughly! Horses, by the way, are not the only species where such breeding schemes are used. Cindy Boulvare, president of the Dallas-Fort Worth Pomeranian Club, talked about Pom color in the *AKC Gazette* (March, 2000): "Black and tan Poms, as well as those with black masks, were not permitted to be shown until 1971, although they were an important part of breeding programs because they provided coat texture."

Presenting only **C** and **c** as allelic choices at the **C** locus may be simplistic. I imagine that the **C** locus in llamas and alpacas features multiple alleles, as found in other species.

I suspect that recessive alleles at the **C** locus are not the only mechanisms that produce phenotypical cremellos. I agree with Dr. Graham that "some llamas are white because they carry genes for multiple diluting genes, possibly including roan. The color of these llamas may show sometimes, and at other times they may appear to be entirely white." In my opinion, this applies to alpacas as well.

Do genes that code for dilution exist in camelids? Here's the tricky part. Dr. Sponenberg observed that there is a "general lack of gold fleece types that have black points" (2001). This throws the existence of at least some dilution genes in camelids in doubt.

Whenever you speculate about the existence of the **C** and **D** loci in camelids, please keep Dr. Sponenberg's reservation in mind. He does not totally rule out their existence, but states that it is "more likely that the fawns are simply Agouti locus alleles, and most breeding results in fact point in that direction."

I will continue to include the **C** and **D** loci in descriptions of possible alpaca color genomes, while remaining aware of Dr. Sponenberg's comments.

For example, my ARI files show the occurrence of two blacks producing light silver-grey as well as that of two light silver-greys producing a black. At first glance, this does not make sense. Is it possible that some of the <u>uniformly</u> appearing light silver-greys are not genetic roans? We may have, similar to the situation of the two "reds," two genetically different light silvers. Double doses of the chinchilla allele (C^+) at the **C** locus as well as the recessive allele (D^+) at the **D** locus have been documented as producing a "silver" coat or fleece in other species.

The allelic choices in this situation would <u>not</u> be black versus silver. Silver does <u>not</u> carry black as the <u>alternate</u> recessive allele. If my hypothesis proves correct, you've essentially bred two blacks. The Punnett Square will help you figure this out: silver ($A^a A^a\ E^+ E^+\ B^B B^B\ C^C C^C\ D^+ D^+$)

bred to a genetically different silver ($A^aA^a\ E^+E^+\ B^BB^B\ C^+C^+\ D^DD^D$) produces black ($A^aA^a\ E^+E^+\ B^BB^B\ C^CC^+\ D^DD^+$).

Don't worry, I didn't dream up these genetic possibilities all by my little old self (okay, maybe I'm not so little).

In *Colour Inheritance in Fancy Mice*, W. MacKintosh Kerr identifies a mouse called the Silver Fox as a^ta^t BB $c^{ch}c^{ch}$ DD PP. In mouse nomenclature, **P** denotes a dark eye and **p** a pink eye. The recessive **p** codes for dilution of coat and eye color. When a Silver Fox, essentially a silvered black and tan, is bred to a pink-eyed silver (**aa BB CC dd pp**), it is fairly easy to figure out that the offspring will be a^ta BB Cc^{ch} Dd Pp — black-eyed black and tans. Breeders expecting silvers from such a mating will be sorely disappointed.

A similar example is the mating of champagne and blue mice (blues can actually look grey). Both are black (**aa**) at the Agouti locus. Because of their other loci — **bb CC DD pp** and **BB CC dd PP** respectively — they produce solid black offspring when mated together. Kerr's very entertaining book (I admit I'm easily entertained) taught me respect and appreciation for the tiny pioneers of color genetics. I no longer complain when my husband does nothing to discourage the presence of these beautiful little creatures in our garage (he claims they clean off the grass underneath the mower's deck).

Our home-bred variety all seem to be the identically dull field brown. Just wait until I spot that first silver critter gnawing away at my canna bulbs!

Kerr's symbols differ from those we've used for camelids. My symbol C^+ for the recessive allele at the **C** locus is the equivalent of Kerr's c^{ch}.

Dr. Sponenberg mentions a possible dilution gene producing a rare charcoal grey in alpacas.

Although we can continue to speculate about dilution genes in the camelid population, the smart breeder will be wise to remember that "dilution effects in alpacas are very rare" (Sponenberg).

Do you find the long rows of letters confusing when planning mating combinations? Use the Punnett Square to work out genetic probabilities — one locus at a time! That will make it easier in the beginning. Eventually, your eye will adjust and you'll take in the "whole picture" at one glance.

The Grey and Roan Loci

Cattle with red coats evenly sprinkled with white hairs are commonly described as *roan*. The choice at the cattle roan locus is **roan - not roan**. In dogs, *grey* refers to a uniformly grey coat. The choice at the canine **G** locus is **grey - not grey**. What about camelids? Explaining, discussing, and reporting on research concerning the inheritance of camelid genes coding for grey or roan is, to put it mildly, one big headache. Why?

Early in the history of the North American camelid industry, one happy and carefree individual decided to borrow the term "grey" and apply it to a roan llama or alpaca. The description caught on fast in the camelid community.

A few authors made valiant efforts to correct this, but owners and breeders in general enthusiastically embraced the genetically incorrect labels. What's more, the colorful critters must be registered as grey rather than roan.

Most people who objected to the popular description finally decided there's no sense in beating a dead horse or (heaven forbid) a living "grey" alpaca or llama. To make sure we are "on the same page," let's review phenotypes and corresponding symbols.

In alpacas, we find two different types of "greys." One is the common black or red animal with more or less evenly distributed mixture of white fiber. Depending on the basic pigment and its modified version, such alpacas are referred to as rose-grey (red) or silver-grey (black). Table C lists the variations. These greys express a white pattern on head and neck that resembles a medium-sized to small tuxedo pattern. Quite often, their legs are white as well.

A BREEDER'S GUIDE TO GENETICS
Relax, It's Not Rocket Science

This is a typical grey alpaca. The head and neck are mostly white. Genetically speaking, this mixture of colored and white fibers should be described as roan (photo by Wool & Gray Alpacas)

In his latest (2003) update on alpaca color inheritance, Dr. Sponenberg assigns G^G to the dominant allele at the Grey locus and G^+ to the recessive. The choices at the alpaca **G** locus are therefore G^G (grey) or G^+ (not grey).

It gets more complicated. There are "greys" without the white tuxedo-like pattern; they have dark heads and legs. Sponenberg believes their mixture of colored and white fiber to be under separate control from the more common greys. To distinguish between the two loci, he names this currently rare pattern *roan*. The choices at the Roan locus are the dominant Rn^R (**roan**) or the recessive Rn^+ (**not roan**).

What happens when a <u>true</u> grey is found in camelids? Maybe the culprit responsible for the current alphabet soup should sit up at night and contemplate that issue (no, I don't know his or her identity, and I don't wanna know).

In any case, the G^G allele coding for grey is believed to be dominant.

If this theory on camelid colors is correct, it would be fruitless to purchase a self-colored animal out of a rose-grey parent and expect it to produce rose-grey offspring with a self-colored mate. You could presumably hope for grey to be present in a very light colored animal, although it may not be detectable to the eye. A phenotypically white alpaca might therefore be a grey. If the red or black pigment has been removed by other genes, you wouldn't know it. Remember, a dominant trait can't skip a generation. As you learned in previous chapters, occasionally the expression of a dominant gene can be suppressed.

It should not surprise anyone who has done his or her genetics home-work that a $G^G G^+$ grey can produce self-colored offspring. In those cases, he or she has passed on the G^+ (non-grey) allele to a $G^+ G^+$ or $G^G G^+$ mate.

Minimally grey animals are not always easy to identify. Breeders can easily miss small grey areas in the fleece, which creates confusion. When a minimally grey animal produces very obviously grey progeny, breeders may mistakenly believe that grey appeared as the result of a recessive gene. Likewise, grey is difficult to see on very light colored animals, and is impossible to identify on white llamas or alpacas.

While the registry, as we discussed, may never reveal all the secrets of color inheritance, it nevertheless serves as a starting point.

Using data efficiently furnished by Alan Schmautz and Dar Wassink (2000), I discovered that as of October 2000, a total of 1,196 crias had been registered in the six grey categories. Since the registry does not differentiate between grey and roan, we can safely assume that roans are included in this data. Two-hundred twenty-one grey crias had <u>both</u> parents registered with colors other than grey. Since white and beige animals effectively hide the

A BREEDER'S GUIDE TO GENETICS
Relax, It's Not Rocket Science

grey pattern, I excluded them from the sample. The results (see Table B) of this file pointed to a recessive genetic mechanism.

TABLE B — Grey Crias (ARI Data, Oct 2000) out of self-colored sires and dams

	TB TB	TB MB	TB DB	TB DF	TB MF	BB MB	BB TB	BB MF	DB DF	DB DB	DB MB
DSG	5	7	0	0	0	1	1	0	0	2	7
MSG	5	14	0	0	1	0	2	0	0	1	6
LSG	3	9	3	1	1	0	0	1	1	1	1
DRG	3	8	0	0	1	2	0	1	1	0	2
MRG	1	9	0	2	2	2	0	1	0	1	3
LRG	0	4	1	0	1	0	0	2	0	0	3
TOTALS	17	51	4	3	6	5	3	5	2	5	22

	DB DF	DB MF	MB MB	MB LB	MB DF	MB MF	DF DF	DF MF	DF LF	MF LF
DSG	0	1	6	1	2	1	0	0	0	0
MSG	0	0	8	0	0	1	0	0	1	0
LSG	0	1	8	0	0	1	0	0	0	0
DRG	0	2	5	0	2	3	0	1	0	0
MRG	0	0	19	0	5	8	0	0	0	0
LRG	0	1	9	0	3	5	1	1	1	1
TOTALS	1	4	55	1	12	19	1	2	2	1

Denise cautioned that I needed to check my theory. My next data request was for the colors of all offspring out of grey-grey matings. Following the laws of genetics, they should have all been greys (roans). This file's arrival promptly deflated and shriveled my theory balloon. Check out Table C.

TABLE C — Result of Grey to Grey matings. The color of 6 other animals could not be determined from the file. Color designations converted to those adopted by ARI on 4/1/1999. (ARI Data, October 2000)

CRIA COLOR	COUNT
True Black (TB)	141
Medium Silver Grey (MSG)	99
Medium Brown (MB)	88
Light Silver Grey (LSG)	78
Dark Silver Grey (DSG)	46
Dark Brown (DB)	26
Light Rose Grey (LRG)	16
Dark Rose Grey (DRG)	14
Dark Fawn (DF)	12
White	7
Light Brown (LB)	5
Medium Fawn (MF)	5
Light Fawn (LF)	1
Beige	1

Well, this is puzzling indeed. What's going on here?

If we subscribe to the theory that grey is dominant (as it is in other species), we can speculate that it is possibly incompletely penetrant in alpacas. In other words, a $G^G G^+$ alpaca may express the $G^+ G^+$ phenotype, but reproduce as $G^G G^+$.

In her correspondence with me, Dr. Graham chose roan to describe the camelid greys. I will not change her nomenclature.

Dr. Graham mentions that roan may vary in its extent (part of the body or all of the body) as well as "have variable expression (a baby may be fully roan, but have no roaning as an adult, or vice versa)." She attributes this to modifier genes and believes that "this would explain why minimally roaned or even apparently non-roan llamas can produce strongly roaned llamas" (from personal correspondence).

Graham's observations seem applicable to alpacas as well. Ironically, while researching and writing this chapter our cria Mendel was born. Mendel's black fleece is minimally grey. The grey pattern appears on the back of his rear legs, along the neck, and in isolated areas on the trunk. How many "Mendels" are registered as simply black or brown — thereby skewing the statistical information I so laboriously gathered? Dr. Graham tells llama owners "to mark down a llama as roan if it has ever shown any roaning at any time of its life, even if it is now solid colored" (from personal correspondence).

I agree with Dr. Graham. Genetically speaking, such a llama is a grey. Likewise, my alpaca Mendel is a grey and should be registered as such. The white animal with a tiny red spot should be registered as a red. Those descriptions would be meaningful to breeders selecting for color. The problem is that most breeders have no knowledge of color inheritance and would be thoroughly confused and irritated by such descriptions. They may even consider them to be fraudulent. I'm afraid we will continue to be stymied by genetically incorrect color registrations until a better system comes along.

Are there true greys as well? The charcoal grey alpaca observed by Dr. Sponenberg may qualify, although such an animal could be a blue dilute instead of a true grey (remember that blue and grey can mimic each other). The fiber color "history" of parents and offspring can be helpful in such a case. Was the baby born black and then changed to charcoal? What color were the sire and dam? What about future offspring? Answers to these questions will help you sort out the genes coding for such a color.

When Barbara Ewing, my canine color inheritance "consultant," took a quick look at the ARI color chart (Table A), she immediately pointed out that the genes coding for dark silver-grey and light silver-grey are probably quite different. The issues go beyond Ewing's observation. Genetically true grey animals in other species are born dark-colored. The dominant **G** allele, such as it exists in horses and dogs, gradually dilutes the basic pigment to the extent that a black or red animal eventually passes as white. The base color of grey foals, for example, actually darkens more extensively before the greying mechanism takes effect (Denise's grey Arabian stallion was fairly dark bay when he was born — only the silvery undercolor of his muzzle indicated to the breeder that he would grey out).

In *Equine Color Genetics*, Dr. Sponenberg quotes the breeders of Percheron horses as commenting that "greys are born black while blacks are born grey." He explains that the latter part of the quote refers to the ashy-grey foal coat of black horses. While the fleeces of llama and alpaca babies can and do undergo subtle changes, none that I am aware of exhibit the dramatic change assigned to the genetic action at the canine or equine **G** locus.

When an alpaca breeder who attended one of my workshops asked me how she could "get a grey" out of her white female, I should have answered: "You can't, because they probably don't exist." However, I knew very well that she was referring to one of the six colors referred to as "greys" on the ARI color chart. The descriptions of rose-greys and silver-greys provide functionally workable labels for their intended purpose — helping breeders identify the appropriate color designations to register their crias. Genetically speaking, their nomenclature is a nightmare and will be more so if we ever discover a true grey gene in the camelid population.

It will be helpful if you think of most of the silvers as black and the rose-greys as red. You should visualize various genes either maintaining the dark, rich pigment or possibly diluting it. Only then should you picture the G locus coding for either G^G or G^+, with G^G "stripping" some fiber of all pigment. Dr. Graham told me that she had "yet to see a roan pattern on black llamas, but I am assuming that this would also result in a silvery grey."

The entire package needs to be viewed and understood as separate loci working together to produce a specific phenotype. There is no single allele that produces, for example, a medium rose-grey fleece on a llama.

All of the greys also express a white tuxedo-type pattern to various degrees. I hypothesized at one time that perhaps the Grey and Tuxedo loci interact in such a way that grey cannot be expressed without the tuxedo gene. The reverse is not true. There are a number of camelids with the tuxedo pattern superimposed over self-colored fleeces.

Dr. Sponenberg discounted my theory of a grey-tuxedo linkage. Why? He did not find any offspring with a tuxedo pattern out of two grey or grey/self-colored parents. Under Mendel's Law of Independent Assortment, such crias would have to be born on a fairly regular basis.

Although the white pattern expressed by the greys is not tuxedo, there is nevertheless a linkage between the pattern and the colored fiber/white fiber mixture seen on the trunk of these animals.

Genetic linkage of color genes does occur in other species. In *Equine Color Genetics*, Dr. Sponenberg describes that in horses "the Tobiano locus, which produces a distinct spotting pattern, is linked to the Roan locus."

White alpacas with an "invisible" grey pattern account for delightful surprises. When your phenotypically self-colored white or beige alpaca produces a "grey" baby with a self-colored mate, don't take this as proof that the grey mechanism is under recessive genetic control. More than likely, the white $G^G G^+$ sire/dam passed on the camouflaged G^G pattern to the cria. A friend recently shared the news that a well-known brown stud sired a rose-grey baby. I'll bet money that the grey pattern came from the cria's blue-eyed white dam. Her sire is a rose-grey. Genetically, she may therefore be a grey herself. It's possible that the stud did not contribute a single DNA sequence to the grey pattern seen in his offspring and was given undeserved credit.

While discussing alpacas, Dr. Sponenberg pointed out that "no greys appear to be homozygous. Breeding records demonstrate that the greys do not have two doses of the roan gene" (from correspondence, 2001). Translated into breeder language: you cannot guarantee that your grey will produce greys, since all are $G^G G^+$. It's the luck of the draw — will G^G or G^+ be passed on?

Remember that the homozygous form of roan is lethal in other species such as the horse. Is it in camelids? At present, scientists have not proven the existence of any lethal color genes in camelids. I've pointed out in a previous chapter that a gene producing a lethal effect in one species may be completely harmless in another. While we should not jump to conclusions, it would be equally foolhardy to assume that lethal genes do not exist in camelids.

Let's go on to spots.

Spotting Loci in Camelids

In the Old Testament, Jacob attributed the appearance of spotted or speckled goats out of self-colored parents to the dam's visual exposure to shoots cut from poplar, almond, and plane trees. Jacob had carved white stripes into these shoots. The clever Jacob had observed that uniformly dark brown or black goats often "brought forth streaked, speckled and spotted kids." ("Genesis 30, Jacob Outwits Laban," *The New American Bible*, 1970.) He did not realize that the carved shoots had absolutely nothing to do with creating the spotted kids. Nevertheless, his intuitive "knowledge" of genetic recessives helped him to enlarge his herd.

Jacob's observations apply to many species — alleles coding for white spots are often under recessive genetic control. This is, however, not universally true, nor are all spots necessarily the result of genetic action at only one locus.

For Icelandic sheep, Susan Mongold tells us, geneticist Stefan Addsteinsson identified "92 different recognized, numbered, and named white markings."

For our purposes, I will restrict my discussion to only four in camelids: appaloosa, piebald, extreme white piebaldism, and tuxedo.

Dr. Koenig differentiates between the S and P loci in her treatment on patterns (spotting) occurring in llamas. In "Inheritance of Fleece Color in the Lama" (*Fiberfest*, Vol. 3, Issue 1, 1996), she lists the Appaloosa locus under S (spots), with the **spotting** allele being recessive to **non-spotting**. She paints a very vivid picture of the genetic mechanism by encouraging the reader to imagine that the **ss** combination "lays a light-colored sheet full of holes over an otherwise colored llama." This is somewhat different from the "stripping" visual, but that's okay. It serves the purpose as well, or maybe better, for that particular locus.

In contrast, Dr. Sponenberg prefers to call such camelids "harlequins," because he sees "important differences from the patterns of the Appaloosa horse" (*ARI Journal*).

Dr. Graham makes a distinction between appaloosa and merle in llamas. She states that red merle is much more common than appaloosa, while blue merle is rare. She emphasized that, unlike in dogs, the homozygous form of merle does not have a deleterious effect on llamas.

Dr. Koenig assigns recessive action to the pattern allele at the P locus, which turns a self-colored llama into a *paint* (piebald, pinto), so **PP** or **Pp= no paint pattern**, while **pp = paint**. I stress again that choice of nomenclature varies from species to species as well as from author to author.

A BREEDER'S GUIDE TO GENETICS
Relax, It's Not Rocket Science

In his workshop notes, Dr. Sponenberg points out that "separate groups of genes control the white markings (i.e., tuxedo front, socks, blazes, etc.)."

Sponenberg differentiates between piebald spotting and tuxedo (caped) spotting (*ARI Journal*, 2001). He defines the tuxedo pattern: "This pattern generally has white on the underline from the head to the rear. Most tuxedo animals have color remaining on the back of the neck, barrel, and back, and are white on the face, lower neck, legs, and belly. This pattern can be super-imposed over any background color, and is probably inherited as a dominant gene."

On a visit to our farm, Dr. Sponenberg pointed to this animal as an example of the tuxedo (cape) pattern. Bred to self-colored studs, this female produced offspring with greatly reduced white areas in their fleeces — see following photo for an example (photo by Harley D. Wood)

The ARI article explains that the tuxedo pattern "can be confused with piebald spotting." Why? The culprits are the modifying genes that lead to what Sponenberg describes as a "wide range of expression." The minimal end of the range of one pattern can mimic the most extensive expression of another pattern. This phenomenon is well established in other species.

A tuxedo front, as Dr. Koenig pointed out years ago in *The Fiberfest*, may be the size of a small scarf. The scarf can shrink to the size of a bow tie.

Based on breeding results observed on my farm, I believe the white markings shown in these photos to be the same pattern — tuxedo — with modifying genes affecting its expression.

I would identify the white areas in this young male alpaca's fleece as a tuxedo pattern. His dam's "cape" has been enlarged; the white area has been reduced to a "scarf." Other breeders may call this pattern piebald or pinto. When discussing camelid patterns, define your choice of nomenclature (photo by Harley D. Wood)

Modifiers play a huge and sometimes fascinating role in the expression of white patterns. Carrying the identical number of modifier genes, male horses and mice express less white than their female counterparts (Dr. Graham, from personal correspondence). She points out that, by paying close attention to breeding records, it is not difficult to change the amount of white in your llama herds — either increasing or decreasing it as you desire. My observations confirm this to be true for alpacas as well.

The term *piebald* is firmly entrenched in camelid breeders' language. I'm not sure about llama breeders, but alpaca breeders often use it indiscriminately for any kind of white pattern. Some camelid breeders believe what they call the piebald pattern (which I would label tuxedo — see first photo) to be recessive. This would make it possible for two self-colored animals to produce such patterned offspring. I have not seen such breeding results. Breeders don't always understand that white animals may express the pattern — you just can't see it.

While I would describe only <u>irregular</u> white areas (without the "cape" or "mantle" effect produced by the colored areas) as piebald spotting, other breeders disagree. Again, minimal expression of tuxedo and piebald may very well mimic each other. If alleles at separate loci code for these patterns, one animal would be able to express both. Phenotypically, the two patterns could merge or overlap and later confuse a breeder with puzzling breeding results.

This is different from canine pattern alleles that reside at one locus (multiple allelism), with the allele coding for self-color being dominant over the pattern alleles. A dog, for example, cannot express Irish markings and a piebald pattern at the same time. He can only be Irish marked and carry a recessive piebald allele (or be homozygously recessive for either pattern).

The camelid piebald (irregular spotting) pattern and the tuxedo pattern must be seen as separate entities for meaningful breeding decisions to be made. The choices at the spotting locus would be **spotting - no spotting**, while the choices at the Tuxedo locus are **tuxedo - no tuxedo**. In addition, we'd have **no appaloosa - appaloosa** at a third locus.

If you recall, I mentioned in Alphabet Soup that piebald technically only refers to a black/white combination. Camelid breeders usually do not make the distinction between piebald and skewbald.

I believe true "piebald" alpacas to be quite rare and have only seen them in photos.

You can see how lack of uniform and consistent identification and labeling of the various patterns makes communication difficult. Camelid breeders must standardize the nomenclature for spotting to have meaningful exchanges of information take place. This issue has been tackled by breeders of other species. Dog breeders, for example, do not usually confuse Irish markings with piebald (irregular) spotting. Communication concerning the alleles coding for these two distinct patterns therefore proceeds smoothly and efficiently. Horses exhibit a variety of spotting patterns. Their breeders dropped the piebald description. "Neither piebald nor skewbald indicates which specific pattern of white is present, and the trend among horse fanciers is no longer to use these terms" (*Equine Color Genetics*). Let's not forget the previously mentioned 92 numbered and named Icelandic sheep markings!

When you discuss pattern genes with other breeders, be sure to define your interpretation of the terms piebald and tuxedo. Is the piebald (irregular spotting) pattern dominant or recessive? I haven't examined any breeding results from animals expressing these patterns.

Armed with information from this book, an interested breeder can study such results and come to his own conclusions. I was told by several breeders that spotting is recessive. With the general confusion covering tuxedo/piebald patterns, I have not put much stock in this information as far as alpacas are concerned.

To further complicate matters, it is possible that camelids carry a recessive as well as a dominant gene for piebald spotting —each residing at a separate locus. Dr. Graham, who admitted to initially arguing against the dominant gene in llamas, eventually came to own a half-Bolivian stud that fits this category. Bred to paint females, this paint male produces solid-colored babies. For the fledgling student of color genes this is nothing to be happy about. It's harder to pinpoint mode of inheritance for individual animals.

Fancy mice have separate genes coding for spotting patterns — one dominant over self-color, the other recessive to it. You see that such a phenomenon in camelids would certainly not be unique among mammals.

An allele for extreme white piebaldism may be present at a Spotting locus (not to be confused with the locus coding for appaloosa). Snowy white animals are produced by two dark-colored parents — both passing on the recessive allele coding for extreme white spotting to their offspring. Remember that dilution genes may produce an identical phenotype.

The Virginal Whites (Some are — not!)

Not all brides who dress in virginal white gowns have "saved themselves" for their intended (nor have all grooms who are decked out in their white tuxedos, for that matter). Likewise, a white alpaca is not always what it shows on the surface. We have discussed white phenotype and genotype in previous sections, so this is just a brief summary. You will recall that Dr. Graham does not believe in the existence of dominant white in llamas. Please remember that skin color needs to be examined, not just the fiber! I know from personal observations what Dr. Graham is referring to. Many blue-eyed white alpacas sport prominent pigmented areas over their body's skin — but their fiber remains snowy white.

Dominant white (Wh^W) cannot skip a generation, and a $Wh^W Wh^W$ or $Wh^W Wh^+$ animal cannot have any colored areas in its fiber or on its skin, no matter how tiny. $Wh^W Wh^W$ animals always produce white in the first generation. $Wh^W Wh^+$ may pass on Wh^+, thereby allowing color to be expressed. The color of $Wh^+ Wh^+$ animals is determined by other loci. The choices at the **W** locus are only **dominant white - not dominant white**.

As discussed, extreme white piebald may or may not exist in alpacas. Good evidence for an extreme white piebald gene is the appearance of all-white babies born to colored parents. The important message here is that dominant whites and extreme white piebalds do not dilute colored fiber to various degrees. It's all or nothing with these guys. Black fiber is not diluted to grey — nor red fiber to fawn — with either genetic mechanism (if it helps, think of them as the *virginal* whites). Close to a century ago, mouse geneticist W.M. Kerr argued with breeders about their mistaken belief that using albino mice would dilute the coat color of their stock.

The third group includes all the "imposters" — the colored alpacas that are phenotypically white because of the combined impact of other genes that remove pigment cells.

Identification of loci and their genetic mechanisms will help breeders to make intelligent decisions on how to add or remove white from their herds.

Dominant white, if it is ever positively identified in camelids, will obviously be the easiest to preserve or to remove. Simply breeding to dominant whites will preserve the trait. To remove it, select the Wh^+Wh^+ offspring from Wh^wWh^+ x Wh^wWh^+ or Wh^wWh^+ x Wh^+Wh^+ matings and you will accomplish your goal. If all Wh^+Wh^+ alpacas are also consistently homozygous for full expression of color at other loci, the probability of white crias being born on your farm is zero. The genotype of such an animal could possibly be Wh^+Wh^+ A^A- E^+E^+ B^BB^B C^CC^C D^DD^D G^+G^+ S^+S^+ Tu^+Tu^+. Alleles at the Agouti locus could vary. I've used S^+ (recessive) to identify self-color.

The extreme white piebald is trickier to eliminate if it is carried as a recessive. Two self-colored animals could still produce a completely white cria if both parents carry one dose of extreme white piebald. Selecting for such whites would be easy: breed extreme white piebald to extreme white piebald.

As in other species, multiple allelism may apply here. The alleles coding for self-color, piebald, and extreme white piebaldism could very well reside at one locus. The individual animal can only carry two.

After studying the action of pigment-removing genes, we know that a red animal, for example, might look white (possibly with blue eyes) if it carries a genotype such as Wh^+Wh^+ A^+A^a E^eE^e B^BB^B C^+C^+ D^DD^D G^GG^+ $Tu^{Tu}Tu^{Tu}$. This phenotypically white individual is genetically a rose-grey animal carrying a recessive for black. It also has dominant alleles coding for tuxedo and grey patterns. The combination of $E^e E^e$, G^G, C^+, and Tu^{Tu} could easily result in almost total loss of pigment. Bred to a black, this alpaca might just surprise you and produce a medium silver-grey.

Remember that white animals can "hide" white patterns. When a white dam mated to a self-colored stud produces a tuxedo-patterned cria, it does <u>not</u> mean that she carries tuxedo as a recessive — she <u>is</u> a tuxedo, you just can't see it. Tricky, huh?

I never heard of an albino camelid. A white alpaca or llama with blue eyes is not an albino.

The Problem of Advertising Color Genes

Let's briefly discuss advertising your stock in the context of color inheritance. A sentence such as "Our black herd sire produces a large number of black crias out of fawn dams" reveals a total lack of understanding of genetic principles on the part of this animal's owner. The gene coding for black behaves no differently for one black alpaca stud than for any of the others.

Let us, once again, use the Punnett Square to clarify a concept.

If a non-black female bred to a black male carries a recessive allele for black, the probability of producing a black cria is 2 out of 4, or 50 percent.

		Sire	
		A^a	A^a
Dam	A^+	A^+A^a	A^+A^a
	A^a	A^aA^a	A^aA^a

If the female carries only A^+ or any of the other agouti alleles dominant to A^a, then a black llama or alpaca stud can't very well sire a black cria out of that particular female.

It's all her fault!

		Sire	
		A^a	A^a
Dam	A^+	A^+A^a	A^+A^a
	A^+	A^+A^a	A^+A^a

If both carry A^aA^a, it's obvious that the offspring will all be black. What about white offspring out of two black parents?

By now you may feel a little insulted that I ask this question — you've learned that such a cria is genetically black (A^aA^a) but had its pigment stripped by other genes.

In any case, the implied promise of a "magic" or "special" genetic color mechanism unique to a particular animal is complete nonsense.

Sometimes advertisements stress that an alpaca comes from a long "line" of animals with black fleeces. I'm not sure I understand the point of such ads. A black alpaca or llama is A^aA^a at the Agouti locus — period! That holds true for a first generation black (out of two fawn A^TA^a parents, for example) as well as for the descendants of 100 generations of black alpacas. <u>The one-hundredth-generation black holds no genetic color advantage over the first generation black</u>.

What about telling prospective buyers that your "grey" alpaca comes from a long "line" of greys, with the implication that this fact is genetically "special"? G^G is dominant — so <u>any</u> grey is descended from a long "line" of greys. It doesn't take a clever breeder to select for a dominant trait, nor does it take a "special" animal to express it.

My advice is to avoid placing advertisements that mention genetics until you feel confident you can make intelligent and accurate statements. Ask or even pay a knowledgeable breeder/geneticist to scrutinize your advertisement to weed out any inaccurate or (genetically speaking) totally bizarre claims.

Souped-up Goats?

Don't think that camelid color genes are necessarily limited to those mentioned in this guide. There may be more loci and/or alleles than I discussed here. The veterinarian who kiddingly (no pun intended) described them as "souped-up goats" might be closer to the truth than we think — at least in the area of color genes.

Well, let's hope not! I have enough trouble sorting out loci with five allelic choices, never mind fifteen or twenty.

Reading any material, even if at this time it includes educated guesses, will be helpful to those breeders who wish to expand their knowledge. As I've already mentioned, Eric Hoffman's new *The Complete Alpaca Book* includes a comprehensive chapter on color genetics contributed by Dr. Sponenberg. Alpaca- as well as llama breeders will greatly profit from reading this comprehensive and authoritative guide.

<u>I emphasize again that my own opinions and speculations expressed here do not present a definitive study of camelid color genetics. My objective is to teach you the genetic "system" so you will understand the information published by scientists and other authorities in the field.</u>

<u>End of Color Chapter, or Why is Denise Smiling?</u>

You have learned how color inheritance in mammals is under genetic control. It is not always easy to understand, even if the process actually evolves in an orderly, well-organized fashion.

Nutrition and/or sun exposure can influence the depth or shade of pigment. However, environment normally does not affect the <u>basic</u> pigment of an animal. There are interesting examples of how seemingly unrelated physical functions impact coat color in some species. One is the dark coloring found on the extremities of Siamese cats and Himalayan rabbits. In these breeds, the biochemical process producing the dark pigment can only work in below-normal body temperatures. The hotter core temperature of the main body does not permit formation of pigment cells.

Scientists have learned that white markings in Holstein cows are not completely determined by heredity. Remember Zita? Her clones are not marked exactly like her. When scientists succeeded in cloning a cat at Texas A & M University, they reported the same phenomenon when the calico kitten was finally born on December 22, 2001. Researcher Dr. Duane Kraemer was quoted as saying: "This is a reproduction, not a resurrection" (Malcolm Ritter, *Associated Press*). Obviously, there's more to the spotting story than only genes coding for such white marks.

There are numerous cases in the animal world where environment can and does impact the expression of certain genes in dramatic ways. Tortoises, for example, are born as females when the eggs have been incubated at high temperatures. Lower temperatures create males. Future research might very well reveal more environmentally sensitive genes in mammals, including those coding for pigment.

Not all breeders may agree with me, but I like to think that Dr. Sponenberg's comment, "A fast Thoroughbred cannot be a bad color" (*Equine Color Genetics*), translates to all species and their specific and unique contributions to the earth's diversity.

A fellow alpaca breeder told me: "I learned from breeding rabbits that you should initially ignore color and visualize each animal as an albino. Choose the most correct specimen — the color is icing on the cake." In the case of lethal color genes or those producing neurological problems, this statement is open to discussion. As a general guideline, it is excellent advice.

Chapter Thirty-one

COLOR-CODING RAINY

CHAPTER THIRTY-ONE

Color-coding Rainy

"Ingrid, you <u>promised</u> the last chapter was the end of the color stuff. I don't know how much more my eyes and fingers can take of this!"
Denise Como

You can "flick switches" and construct an entire color code for each individual animal as long as you are familiar with colors and patterns going back several generations in the animal's pedigree. Let's use my Whippet bitch Rainy as an example.

As mentioned earlier, Rainy is a red brindle with a black mask and a dark belly. She carries tiny white marks on her chest and under her chin. Her sire is self-black, her dam is dilute red brindle with piebald spotting. The dam most certainly carries and expresses one allele for chinchilla dilution, making her red coat look tan. On Rainy's maternal side, her grandparents are both brindles — one with piebald spotting, the other a dilute brindle with no mark-ings. On the paternal side, her granddam is black, and her grandsire is brindle with piebald spotting.

Starting with the locus for dominant white and methodically working our way through all loci, we finally arrive at the following code: **ww $a^y a^y$ BB CC DD $E^m e^{br}$ Ss^p gg mm rr TT.**

Let's analyze: there are no dominant white dogs (even if there were, Rainy would not qualify), so Rainy is **ww** at the **W** locus. She is not black, so she cannot be either **AA** or **A** in combination with other alleles. Both her parents contributed red (the dam is $a^y\ a^y$ herself, the sire is **Aa^y**), thus we must assign $a^y\ a^y$ to their daughter.

Could Rainy possibly be **Aa^y ee**, the "other" red?

If the **e** allele exists in Whippets, it is extremely rare. Most red Whippets are $a^y a^y$. Red Chow Chows and Greyhounds, for example, can either be $a^y a^y$ **EE** or **AA ee** — with $a^y a^y$ **Ee** and **Aa^y ee** presenting two

253

other combinations coding for red. Remember when I told you that two genetically different red dogs can produce black? After reading the color chapters, you should now understand that remark. Just in case you're having a dense moment (I have lots of them when I read books on genetics), let me help you out with our trusty Punnett Square. We're breeding $a^y a^y$ Ee (red) to Aa^y ee (red). The latter is actually a "disguised" black.

		Sire	
		$a^y E$	$a^y e$
Dam	Ae	$Aa^y Ee$	$Aa^y ee$
	$a^y e$	$a^y a^y Ee$	$a^y a^y ee$

Mendel's Law of Independent Assortment is at work! The Greyhound puppy in the top lefthand corner is black — Aa^y Ee.

I digress — which is so easy to do with this subject matter. Let's get back to Rainy!

The $a^y a^y$ dogs are born very, very dark colored, with an even darker dorsal stripe (dorsal = along the back). Rainy looked almost black at birth, and is a good example of how such animals lighten quickly within a few days of drawing their first breath.

We can be confident that Rainy is $a^y a^y$.

The liver (b) allele does not exist in Whippets, so Rainy must be BB at the B locus. All black pigment (nose, mask, brindling) therefore remains black, and is not modified to liver. Unlike her dam's diluted red, Rainy's color is rich and dark, making her CC and DD. E^m assigns her a black mask and e^{br} assigns the brindle striping, another clue that she does not carry ee.

Since Rainy is not grey, gg applies. The merling pattern does not appear in Whippets, so the only choice at that locus is mm. I include rr (no roan) for the sake of a complete profile. Aside from the tiny white marks on her chin, chest, and feet — which often occur in otherwise self-colored or solid brindle dogs — Rainy does not show any spotting. Since her dam is $s^p s^p$, she must carry one piebald allele, making her **Ssp**. Rainy does not show any ticking. Nevertheless, I feel justified in assigning **TT** to her. I base this conclusion on my knowledge of her pedigree, and the heavy ticking exhibited by all her siblings (with the exception of her phenotypical "twin" brother, Lightning). Their dark base color probably hides the dark spots on their skin.

"Hey," you say, "let's go back to $E^m e^{br}$." Okay, if that combination doesn't make sense to you, pat yourself on the back. You are very astute! "How," you ask, "can brindle be expressed, when it is recessive to E^m, meaning E^m should suppress the expression of e^{br}?" Dr. Willis states that "$E^m e^{br}$ will have a black mask and will be brindle in the tan regions."

It was Dr. Clarence Little who first allowed that, in the case of multiple alleles, it is quite possible for two alleles to be expressed simultaneously. In *The Inheritance of Coat Color in Dogs* (1957, 1984) Little describes e^{br} as "... clearly epistatic to **e**...," but "... often imperfectly and incompletely hypostatic to **E**." One can assume that this applies to E^m as well, thus the existence of the distinctly black-masked and clearly brindled Rainy. This would also mean that no dog can be homozygous for black-masked brindle, meaning this particular combination <u>can never be passed on by one parent as a complete package</u>.

Please note Dr. Little's choice of the terms *epistatic* and *hypostatic*. The scientific community no longer uses epistatic and hypostatic to explain the relationship between alleles at loci having more than two (multiple allelism). Of course, Dr. Little died quite a few years ago, so we don't know what terminology he'd prefer now. In any case, keep going — eventually, it will all become clear.

As I mentioned earlier, there is another viewpoint: long-time Borzoi breeder and AKC judge Barbara Ewing feels strongly that the brindling trait is not part of the Extension locus at all. She speculates that it is carried on a separate locus. Ewing comments that "brindling could be just a way of 'organizing' the black pigment already present in the coat through the expression of a^y and **E**."

After hearing Ewing's opinion, I find it interesting that Dr. Sponenberg, when discussing the rare brindling pattern in horses, tells us that "brindle seems to require sooty black countershading for its expression, and reorganizes sootiness into vertical stripes instead of a more uniform sprinkling of hairs."

Dr. Willis, when describing the Agouti series, writes "some breeders of Pekingese call sables brindle but this is erroneous nomenclature." Hmmm — maybe those old Pekingese breeders were on to something after all!

Sue Ann Bowling, a guest columnist in the *AKC Gazette* magazine (October 1999), tells Shetland Sheepdog breeders: "The brindle pattern is dominant and should not exist in the breed today. A brindle puppy did show up at a pet store recently, and I know of one dark sable that was wrongly considered a brindle by a judge not familiar with Sheltie colors."

By the way, many Whippet breeders are quite confused about the relationship between black and brindle. The choice is not black or brindle. It's **black** or **other Agouti alleles** at one locus, and **brindle** or **not brindle** at another locus. A dog could be **AA $E^m e^{br}$**. You wouldn't see the brindle pattern because of the otherwise black coat. Black is not dominant over brindle — two separate loci are involved! If the theory of a separate locus for brindle as proposed by some breeders is correct, we'd count three loci (the genotype would then read **AA $E^m E^m$ Br Br**). If dominant black is not part of the Agouti locus, we'd have to count four. In any case, a dog can be black and brindle at the same time — with the black coat concealing the brindle pattern.

Please don't think that my presentation in this modest guide encompasses the entire color story. It doesn't begin to cover the subject — not even for dog breeders. If you are truly interested in that facet of breeding, you need to move beyond the beginner's guide scope of this book.

The most important message I'd like you to come away with after reading the last few chapters is the fact that things are not always what they appear to be on the surface. As a raw novice student of color genetics, I first assumed the entire subject could be neatly divided into dominant and recessive genes. As we all know by now, the issue is quite a bit more complicated than that.

It was a shock to my tender beginner's psyche when Barbara Ewing mentioned to me years ago that yellow Labrador Retrievers are genetically black! It's not that I doubted her — it just seemed entirely too weird to understand. Dr. Sponenberg's explanations of how chestnut color is formed in horses finally clarified the Lab mystery for me.

I recently read a definition for the word *heterozygous*. The author used color genes and told readers that white is recessive to black. The implication was that the genes coding for white and black fiber reside at one locus. We know they don't. Breeders must see beyond the simple dominant-recessive relationship at only one locus if they want to truly understand the inheritance of genes coding for color and patterns.

Once again, the purpose of my writing was to explain the basic "system" to you. Hopefully, you will now understand in broad terms the research published by scientists who have studied the subject in depth. The inheritance of color genes as a science was firmly established decades ago. Many breeders have a passionate interest in this subject, and they will be happy to share their knowledge. Remember to stay open-minded to new research and discoveries in this complex and fascinating field.

There is much breed-specific information available to interested novices from various sources. Its scope varies greatly from one breed to the next. You will read conflicting information, including theories that will differ from those proposed by Dr. Little and Dr. Willis. You will now be prepared to weigh the evidence and arrive at your own conclusions.

Chapter Thirty-two

BLUE-EYED GIRL

CHAPTER THIRTY-TWO

Blue-eyed Girl

"It is probably not possible to completely avoid all blue-eyed animals, and multiple genetic mechanisms probably account for the different types of eye color."
D. Phillip Sponenberg D.V.M., Ph.D., *The Alpaca Registry Journal*, Spring 2001

"Oh, look at the beautiful blue eyes. Isn't she gorgeous?" "How stunning. Do most alpacas have blue eyes?" "She looks like a person with those eyes. I love it!" The comments made by the general public about Kalita, our blue-eyed girl, have been nothing but enthusiastic. One noteworthy exception was from acquaintances of ours — obviously not animal lovers. Gazing briefly at our freshly shorn Huacaya alpacas, they felt that the members of our little herd looked like "ugly Dr. Seuss characters, especially that blue-eyed one." My husband's response, thankfully uttered in private after their departure, cannot be printed.

Like all other species, alpacas suffer from an array of congenital defects, some of them life threatening. Although many anomalies are suspected to be of genetic origin, this has not been documented at present. None seem to stir up a debate as heated as those "baby blues." Why is that? What makes people who delight in the birth of a blue-eyed child object to the sight of an alpaca with identically colored irises? Does the blue-eyed trait cause negative physical repercussions for alpacas or other species with that phenotype (blue eyes, white coat or fleece)?

Let's back up and explore the general concept of eye color. Multiple genes (polygenes) contribute to eye "color." Alleles that code for *pigment* are usually (but not always) dominant over those coding for *lack of pigment*. A specific locus may be a major contributor. In horses, as we discussed earlier, a double dose of the recessive **c** turns a dark-eyed chestnut horse into a blue-eyed cremello. Whether one locus or several loci are involved, the "blue" we see in the eyes of white or cream animals is not a color but rather a lack of melanin in the eye's iris. The "blue" color is an optical illusion, such as we experience when looking at the "blue" sea.

The chemical process for the formation of melanin is no mystery. The amino acid tyrosine (tyr)-UAU, UAC- is converted to dopa and oxidized to dopa quinone. From there, "dopa quinone molecules undergo spontaneous polymerization to form melanin" (Robert C. King, *A Dictionary of Genetics*, 1974). Lack of pigment results from a deficiency of *tyrosinase*, the enzyme responsible for the conversion. The chemists among you can find a segment of melanin's chemical structure on page 177 of King's book.

The eye "color" (actually a reduction of pigment) in itself does not create problems, but the genetic mechanism — read chemical process — that strips or dilutes pigment from the animal's body *can* cause deafness.

The occurrence of such a defect is not uncommon, and information pertaining to it in other species is not difficult to find.

In *The Merck Veterinary Manual* (Eighth Edition, 1998, pg 369) we find the following explanation for cats: "An autosomal gene in cats causes white fur, blue eyes, and deafness; it is dominant with complete expression for white fur and incomplete expression for blue eyes and deafness. Deafness in this instance is due to cochleosaccular degeneration changes that are expressed in the first week of life." The last sentence I quoted tells us that such cats are <u>not born</u> deaf.

The mode of inheritance for white fleece and blue eyes is obviously not always — and possibly never will be — the same for alpacas as it is for the white-furred cats. None of the white, blue-eyed alpacas I am personally familiar with have a dominant white parent, like the cats mentioned in the Merck text. That fact is immaterial to the argument I will make here.

As mentioned earlier, even if camelids do not carry the dominant allele at the **W** locus, they nevertheless carry other genes coding for lack of pigment. We've discussed possible "whitening" genes in previous chapters.

The **C** locus may be the cause of "true" cremellos in alpacas (as in horses), but it seems logical and possible that other loci might be involved as well. I've also mentioned that the <u>combined</u> effects of two or more dilution and pattern genes at, for example, the **E, D, S, G,** and **R** loci could very well produce blue-eyed white animals, and possibly result in a phenotypically "cremello" alpaca.

Breeders need to look beyond the genetic model of a single recessive gene coding for blue eyes. The Grey locus allele, producing rose- and silver greys, is often implicated in the expression of blue eyes. It is believed to be under dominant genetic control. Intelligent dialogue about the genetic origin of blue eyes requires knowledge of <u>fleece color</u> inheritance.

What is *cochleosaccular degeneration*? Dr. George M. Strain (Louisiana State University Office of Research and Graduate Studies, and School of Veterinary Medicine) tells us on his web site that "the deafness, which usually develops in the first few weeks after birth while the ear canal is still closed, usually results from the degeneration of part of the blood supply to the *cochlea* (the *stria vascularis*). The nerve cells of the cochlea subsequently die and permanent deafness results. The cause of the vascular degeneration is not known, but it appears to be associated with the absence of pigment producing cells (*melanocytes*) in the blood vessels."

Dr. Strain writes: "Blue eyes, resulting from an absence of pigment in the iris, is common with pigment-associated deafness but is not, in and of itself, an indication of deafness, or the presence of a deafness gene." (Millions of blue-eyed Germans, Scandinavians, and Dutch, as well as my red-headed, blue-eyed husband, will be relieved to hear this.)

Australian Shepherd breeder and llama owner Pam Bethurum, writing for *Aussie Times* (September-October, 1998), discusses how the *organ of Corti* (sounds like a vacation spot!), located within the membrane of the cochlea, contains hair cells which act as receptors for air vibrations (sound). Bethurum continues: "The hair cells convert the mechanical energy to electrical nerve energy which passes along the cochlear nerve and on through the network of nerves to the brain stem. This is where the lack of pigment comes into play in causing deafness. In order for the hair cell to convert the mechanical energy into electrical or nervous energy, the hair cell must contain a pigment cell."

In any case, scientists as well as knowledgeable, experienced breeders implicate lack of pigment producing cells as the cause of deafness in animals of various species. Many dog breeders submit their animals to BAER (Brainstem Auditory Evoked Response) testing to detect hearing problems in their lines. A correlation between lack of pigment and deafness cannot be denied.

Several alpaca breeders have provided me with thought-provoking information. Some reported that their cremello alpacas could hear perfectly well, others were convinced that their animals suffered from partial or complete hearing loss. All reported that colored (!) crias out of cremello dams heard as well as any other colored alpacas. This gives credence to my theory that cochleosaccular degeneration as recognized in other species may apply to alpacas as well. The defect is, in my opinion, caused by the loss of melanin, and is not the result of a defective piece of "equipment" in the inner ear.

I doubt that deafness in phenotypical cremellos is caused by a mysterious, defective gene that will be "discovered" by a scientist one memorable day in the future. I don't believe that a genetic marker leading to a deafness gene <u>unrelated</u> to lack of melanin will ever be found in cremellos (deafness in dark self-colored alpacas would, of course, be a completely different story). I think that only a comprehensive study linked to the inheritance of <u>fleece color</u> as has been conducted for the coat colors of other species (dogs, for example) will shed total light on the matter.

Dr. David Anderson of Ohio State University graciously agreed to discuss this issue with me when I phoned him in the Spring of 2000. He readily allowed that my theory as applied to alpacas was a distinct possibility. As a true scientist, however, he advised caution. Anderson would not speak in the affirmative until research confirms such a theory. He felt that his study conducted at OSU, limited in scope due to meager funding, has not been extensive enough to justify reaching any definitive conclusions.

The December 2000 issue of the *GALA Newsletter* featured an article on the subject of congenital deafness in camelids. Co-authors Philip A. March, D.V.M, M.S., and David E. Anderson, D.V.M., M.S., wrote that congenital deafness "is prevalent in camelids with a white hair coat and blue eye color." Their research at Ohio State University proved 90 percent of such animals to be deaf. They further found "one solid color-eyed and white-coated female with deafness, and one solid sky-blue eye with red coat with normal hearing."

Using the BAER test, March and Andersonconcluded "congenital deafness in camelids appears to be due to premature degeneration of sensorineural structures early in life…"

The alpaca with the red fiber/sky-blue eyes combination tested out to have normal hearing. And why shouldn't it? In this case, pigment-filled hair cells did their job — giving the animal normal hearing.

"True" blue eyes, as opposed to *glass eyes* (those lacking pigment), are known in other species as well. Dr. Sponenberg tells readers in *Equine Color Genetics* that "the genetic mechanism behind these is very poorly understood."

Combinations of one brown eye and one blue eye further complicate matters.

Your typical blue-eyed white alpaca is not an albino. The albino gene is usually inherited as an autosomal recessive gene. As discussed previously, it does not dilute pigment to various degrees but offers a very simple **albino - no albino** choice. Albinos don't have blue eyes, anyway.

Why can some of the blue-eyed whites hear?

The issue is not fiber color alone. Some white-fibered animals have dark skin (sometimes only in patches). Those individuals could possibly have enough pigment cells in the ear to keep the hair cells which act as receptors functioning properly. If you recall, in cats the gene coding for blue eye-"color"/defect shows incomplete penetrance (*The Merck Veterinary Manual* calls it *incomplete expression*). Such a mode of inheritance may apply to alpacas as well.

A minimum of pigment does not automatically assure perfect hearing — as many dog breeders can testify. One dark-eyed white alpaca in the OSU study tested deaf. There are possibly others.

In any case, breeders are kidding themselves if they believe the cremello-deafness connection does not exist or occurs only rarely.

During our conversation, Dr. Anderson questioned the ethics of breeding any animal with a defect. Defects are numerous in all species, so breeders of <u>any</u> species must examine the issues and determine selection criteria for their programs.

Since the total elimination of all defects remains a fantasy, choices have to be made.

Which criteria will you select for the elimination of defects, and how rigidly will you apply them? Criteria will vary from species to species and the purpose for which they are bred. While a deaf alpaca or sheep can function very well as part of a herd (and I bet the defect goes undetected many times because of this), a deaf dog will present certain problems. Its handicap must be taken into consideration while training it to become a member of your family. I imagine that deafness could also be a problem for pack llamas and those expected to participate in pulling carts. Alpaca breeders who wish to clicker train all their animals (maybe with the focus of their breeding program on the pet market) may need to evaluate their selection process in regard to alpacas with all or predominantly white fiber.

Keeping in mind that no population can be cleared of all defects, you should ask yourself these questions:

Is producing a deaf cremello worse than an alpaca with a hernia, which needs to be surgically repaired? Is it worse than the occasional cryptorchid in your line, whose retained testicle must be removed to prevent it from becoming cancerous? Does it cause the same hassle as malocclusion (a misaligned jaw), which necessitates the floating (filing) of teeth? Does it cost you precious hours of your time, like bottle-feeding crias whose moms never have enough milk? Must you choose euthanasia or a huge veterinary bill as you do for those animals suffering from choanal artresia? Will it need to be treated with special salves and potions like those animals afflicted with alopecia (loss of fiber)? Does it suddenly drop dead in its pasture like the cria that inherited just a few too many polygenes coding for a heart defect? What *are* the ramifications of maintaining a deaf herd animal bred to produce fiber?

Should we look down the road to the time when alpacas will be grazed, like sheep, on open, unfenced land? Will deafness be a liability then? Apparently it does not pose a problem in South America, but conditions there are not identical to those found on the North American continent.

Should animals with poor conformation (and there are plenty of those!) be selected for breeding, while you cull your perfectly conformed blue-eyed white alpaca from your breeding program? Should a show judge arbitrarily decide, without valid testing procedures in place, that all blue-eyed white animals are deaf and therefore not worthy of consideration to place high in a large exhibition class?

Tough questions, with no easy answers! It all comes down to priorities, compromises, and choices.

Should alpaca breeders encourage and advocate the preservation and registration of a wide variety of colors and patterns, yet look disdainfully <u>at the extreme product of such dilution and pattern combinations</u>? This reminds me of dog breeders who, in total ignorance of genetics, insist on trying to breed for black noses on liver-colored dogs. It is genetically impossible.

Are breeders ready to discard the many beautiful and exotic color variations and patterns that make this camelid breed so unique? Think twice before discarding genetic material that cannot be retrieved once those genes are lost.

It is clear that rigidly selecting against blue eyes translates into selecting against any genes removing pigment. Some parent clubs of AKC dog breeds have done just that: white, cream, fawn, blue, and grey animals may not be registered. The choices in some breeds are mostly restricted to black, red, and the original (wild type) agouti color — accompanied by dark eyes. Several horse registries restrict color to specific choices.

Should breeding of registered alpacas be limited to only medium to dark, self-colored animals to minimize the possibility of producing a cremello?

While I personally like the various shades of medium to dark brown, I would not want to see world-wide herds of uniformly brown alpacas. How boring!

Steady, relentless selection pressure for exclusively dark colors would <u>eventually</u> eliminate the occurrence of blue-eyed animals. Is that what the alpaca industry wants to achieve?

So far, the choice of selecting for or against genes coding for depigmentation (read: lack of melanin) rests with individual breeders. The decision to completely eliminate the cremello phenotype from the alpaca gene pool should only be made based on in-depth knowledge and expertise in the field of color genetics, with <u>full comprehension of the possible repercussions</u>. Enough damage has been done to various animal populations over the years by breeders who culled all breeding stock with certain traits, based on poorly understood genetic principles.

I respect and appreciate the right of the individual breeder to make a choice. After weighing the issues, I can see valid arguments made for both sides. However, if you call for the elimination of all blue-eyed whites, don't be surprised if I question the presence of beiges, greys (roans), and dark-eyed whites in your pastures — all phenotypes that genetically contribute to the appearance of cremellos in our herds.

Many breeders have expressed confusion over the advice to breed cremello females but to geld males. No wonder they are confused. There is no scientific evidence to validate that advice. With rare exceptions, genes coding for colors and patterns are not sex-linked and therefore are not carried on either the X or Y chromosome. Breeders who believe that males are always more "important" need to revisit the information on inheritance of genes (including mitochondrial DNA!). They should re-read the section about the use of <u>herd sires</u> (males covering an entire herd) versus <u>studs</u> (males bred selectively to hand-picked females).

The choice to use a cremello male should rest on sound knowledge of genetic principles. Your own personal breeding objectives play a role here.

There are programs where the use of such an animal would be extremely foolish. There are others where the <u>selective</u> and <u>informed</u> use of a cremello alpaca <u>stud</u> can be as justified (or not, depending on where you stand on this issue) as using a blue-eyed female.

Breeding a poor or mediocre blue-eyed female — but gelding an out-standing blue-eyed male (as I observed at one farm) — does not make sense, genetically speaking.

The wonderful aspects of learning about genetics is that accumulated knowledge can be <u>applied</u> in a very practical manner, as well as vigorously argued over with other breeders.

Let's not forget that deafness can have other causes, environmental as well as genetic. A German newspaper, for example, reported that South

American breeders are experiencing problems with their alpacas becoming deaf after repeated pesticide baths to treat for external parasites.

For an interesting presentation on the effects of depigmentation on the general nervous system, as well as defects and behavior patterns associated with animals lacking pigment, you may want to read *Genetics and the Behavior of Domestic Animals* (Temple Grandin and Mark J. Deesing, 1998).

Chapter Thirty-three

COMPROMISES

CHAPTER THIRTY-THREE

Compromises

"Compromise: an adjustment of opposing principles, systems, etc., in which part of each is given up."
Webster's New World Dictionary

Occasionally, I envy those breeders whose knowledge of genetics can fit in a thimble. Such people go happily about their lives, unencumbered by complex decision-making processes.

However, I also think of how much fun they're missing. So, okay, a breeder's idea of fun may differ vastly from that of the rest of the population. "Fun" may include conversations over dinner about subjects like parasites, emptying anal glands, and how your toddler ate a cup of kibble put down for your old brood bitch (my son Benjamin called it dog candy). My non-farming family members in Germany worry about what they consider my "strange" preoccupation with testicles — which stems from the fact that cryptorchidism is a defect common to Whippet lines. What would my family say if I, like Mary Fricke (*Dino Godzilla and the Pigs*, 1993), lived on a Missouri pig farm and served mountain oysters (fried hog testicles) at parties?

Please note: "Dog breeders generally refer to males with only one descended testicle as being monorchids and those with none descended as cryptorchids. In this connection, breeders are using faulty definitions. A dog which has undescended testicles is a cryptorchid and may be unilateral (one descended) or bilateral (neither descended), whereas the true monorchid only actually has one testicle in the body which may or may not be descended" (Willis, 1989).

So far, all my "boys" are batting 100 percent, although a bitch I bred and co-own produced a unilateral cryptorchid. Once I actually sold a little guy as a cryptorchid with the agreement that he would be neutered. When the call came weeks later that Woody's (the owners named him after me) second testicle had "dropped" (breeder's jargon for a testicle descending into the scrotum), the joyful news was known to at least five other Whippet households on the Eastern seaboard within the hour. So pronounce me nuts (pun intended); we breeders take that stuff seriously, and well we should.

Anybody checking out a stud should request to see his testicles. If this embarrasses you, practice asking the question aloud at home first. (I'm not suggesting, however, using the postman or the washing machine repairman for practice!) If the prospective stud has only one testicle, or they're the size of small marbles (*testicular hypoplasia*), you should re-think your choice.

The jury is still out on the exact genetic mechanisms behind cryptorchidism or monorchidism, though many breeders consider them recessively carried faults. With other choices available, there is no reason to use a stud showing such a defect. Scientists have positively identified a correlation between size of testicles and fertility in livestock. In dogs, "small, soft testicles are usually associated with poor semen quality..." (*The Merck Veterinary Manual*). Don't count on show judges to disqualify defective animals, and don't take the stud owner's word that his stud's equipment is in A-1 condition.

The study of genetics is so utterly fascinating and multi-faceted that a person could fill a lifetime pursuing it and never learn enough. I find it intellectually stimulating and richly gratifying.

One of the most helpful personality traits to a breeder is the ability to arrive at compromises. Webster's Dictionary defines *compromise* as "an adjustment of opposing principles, systems, etc., in which part of each is given up." As your understanding of genetics grows, you realize that each breeding is exactly that: a compromise. Anyone who tells working mothers and breeders that they "can have it all" is a bald-faced liar. Remember that! Except in the eyes of some raw and naive novice breeders, no one animal achieves total perfection, no sire or dam adds everything you need to your program. Phenotypically "perfect" specimens all have skeletons hidden in their genetic closets.

The need to compromise does not entitle a breeder to practice deception or smudge the truth. The bag of tricks used by unscrupulous breeders of all species to conceal faults and defects is filled to overflowing. Investigate carefully, so you are not left "holding the bag." Enough honest, ethical breeders exist to offer you choices. Do not give your business to those who have neither your interests nor those of the animals they breed at heart.

Novice exhibitors and breeders rarely have a clue about the extremes some slick and callous breeders will resort to in order to win in the show ring. In case you feel I am vastly exaggerating, please read on. In June 1999, AKC Board Members voted to suspend for four years a dog owner from Canton, Ohio "for exhibiting, or causing to be exhibited, a male Mastiff at eight events which she knew, or should have known, was ineligible to compete because it had been changed in appearance by artificial means as a result of the insertion of an artificial testicle…" (*AKC Gazette*, Secretary's Page, October 1999).

This is not an isolated incident by any means. Misaligned teeth are routinely "fixed," missing pigment filled in either with permanent ink pens or tattooing, greying muzzles dyed with black hair coloring, cigarette ashes used to darken light areas just before show ring time, "bad" tails broken and reset, thin wires placed in ears to attain the "correct" ear carriage — the list is horrific and endless, and this is just in dogs!

Now you know! When contemplating the purchase or stud service of an animal, not only do you have to speculate upon the genotype not expressed by the phenotype, you must also ask yourself, "Am I seeing the true phenotype or the clever work of a surgeon and/or groomer?" Not only should you ask to see, you should also insist on touching.

Denise related a story to me about finding a stud dog that she felt was the perfect match for one of her Borzoi bitches. Photos, advertisements, pedigree, a glowing description by the owner — the dog was ideal. She was so excited when she knew she was going to see the stud "in the fur." When she finally got her hands on the dog, she found that lots of coat and an excellent grooming job cleverly hid a serious lack of substance, poor structure, and no muscle tone. She was sorely disappointed and vowed never again to attempt to make a breeding decision based on photos and word-of-mouth.

If the frequent mention of defects has worried our neophyte breeders, I apologize. No genetic text, however basic, can be written without broaching the subject. It must be dealt with — without pointing fingers, spreading rumors, or making caustic remarks — by way of a positive outlook and a realistic approach towards the common goal of eliminating as many problems as possible.

Many breeders paint, sculpt, write, weave, spin — they are creative individuals. Breeding animals also satisfies the urge to create something unique and wonderful.

We should not lose sight of the fact that animals, unlike paintings or sculptures, are living creatures. We share much of our genome with them (close to an amazing 99 percent in the case of the chimpanzee) and have the responsibility to approach breeding any species with as much knowledge, wisdom, and decency as possible. Animals have served mankind well for millennia, and they continue to do so in increasingly complex ways. They are living beings — they are entitled to be treated with thoughtful kindness.

> *I am the voice of the voiceless;*
> *Through me the dumb shall speak,*
> *Till the deaf world's ear be made to hear*
> *The wrongs of the wordless weak.*
> *And I am my brother's keeper*
> *And I will fight his fight;*
> *And speak the word for beast and bird*
> *Till the world shall set things right.*

>> Ella Wheeler Wilcox (1850 - 1919)
>> As published in the *Gala Newsletter*,
>> Volume XV, Number 1, February 2000

Chapter Thirty-four

SURI I AND SURI II

CHAPTER THIRTY-FOUR

An Introduction to Suri I and Suri II

I vacillated between including this two-part series or leaving it out entirely, as it has been featured in several camelid publications. After much thought, I decided to present it with minor changes and additions.

This is not a hypothetical situation but a real-life scenario that is frequently encountered by alpaca breeders. Its contemplation presents an excellent opportunity to demonstrate the complexity of genetic issues. Additionally, I am positive that parallels can be drawn to similar situations facing breeders of other species.

Finally, I cannot stress enough how much information the study of all creatures has to offer to those breeders thirsty for knowledge. More than likely, my little "zoo" at Stormwind Farm will continue to include only Whippets and alpacas. This will not curtail the pleasure I experience from studying other breeds of dogs and livestock species in any way. I sincerely hope you share that feeling.

Suri Genetics — Part I

Alpacas come in two varieties — Huacayas and Suris. Huacaya fiber "has a fluffy, spongy appearance" (*The Alpaca Book*, Eric Hoffman and Murray E. Fowler, D.V.M.). Suri fiber has more luster and "grows parallel to the body, often hanging in curly ringlets." Suris are more rare than their teddy-bear-like Huacaya relatives. Both imported and American-bred Suris do not always breed true to type, but frequently produce Huacaya crias.

Some breeders passionately believe in setting purification of the Suri type as a goal for his or her breeding programs.

Others have taken the stance that Suri-Huacaya crosses benefit the Suri population by adding hybrid vigor and additional color genes to their mostly white Suri herds.

That the continued use of colored Huacayas adds color to your Suri pastures is a given. Granted, not every cross would result in a colored cria, but a sufficient number would be produced to satisfy most breeders.

The argument of hybrid vigor as it pertains to Suri-Huacaya crosses is a specious one at best. The concept of crossbreeding followed by backcrossing into the original breed or line is not a particularly revolutionary one. Its practice is widely discussed among serious breeders of livestock and purebred dogs. The use of this system had and continues to have practical implications for various species. For example, breeders have crossed two (or more) separate and distinctly different breeds to create a new one. Whippets, to name one, were deliberately created from greyhounds and various terrier breeds to fulfill a specific purpose. Irish Wolfhound breeders used Scottish Deerhounds (and possibly other breeds) to re-vitalize a breed that was in danger of becoming extinct. They were successful in this undertaking. The modern Irish Wolfhound cannot be distinguished from its ancient predecessor, although the show ring added bulk and weight that the early cousins probably didn't have. This particular genetic model does not apply to Suri-Huacaya crosses.

While Suris and Huacayas sport distinctly different fleeces, they are no different than the "powder puff" puppies born in Mexican Hairless and Chinese Crested litters. Long-coated and smooth Saint Bernards are not two separate breeds — merely two distinct coat varieties. Likewise, the Suri male standing at stud on a farm three states over could conceivably be genetically closer to your Huacaya import than the Huacaya stud at the farm three miles down the road. Hybrid vigor does not come into play under such circumstances.

The livestock industry also uses crossbreeding to produce commercial market animals, taking advantage of the hybrid vigor created by such breedings. These crosses ultimately end up on your dinner table or on your feet. They are not routinely used to pursue a superior end product that spans many generations. They *are* the end product! Desirable as well as undesirable results of such crossings simply disappear.

A look in the farm listings of the AOBA Breeder's Directory shows us that the majority of alpaca males are not even gelded (although many geldings are probably not registered). To my knowledge, very few breeders remove large numbers of females from their breeding programs (I am not passing judgement here, just simply stating a fact).

There exists a third purpose for crossbreeding. In this case, it is more commonly referred to as outcrossing. Individual geneticists sometimes use different nomenclature for breeding systems. Most breeders use the term outcrossing for "the mating of unrelated animals of which one is or both are inbred [or linebred]" (Malcolm B. Willis, 1989).

From the pedigrees I have seen I suspect that the majority of North American breeders have not used a systematic program leading to a line, but are choosing studs based on phenotype alone. There is nothing wrong with such a program, and it can stand on its own merit. It does not, however, lead to the formation of a distinct line. Breeders of purebred as opposed to crossbred commercial stock (as in crossing members of two distinct breeds) as well as companion animals, often take great pride in having established a distinct line of animals. Offspring bred by such people carry and pass on a certain phenotype (appearance) that is easily recognized by others knowledgeable in the breed. They are all very uniform in conformation, size, and possibly color.

The show ring encourages such trademark "looks," for better or for worse. How do you achieve such uniformity? Breeding animals with similar phenotype is one way, but it is often not very effective. The quickest and most efficient way to reach that goal is through inbreeding (followed by rigorous culling of undesirable phenotypes) or intense linebreeding. In- or linebreeding can produce such prepotent animals that ten generations later some of their descendants are still recognized as "going back to old so-and-so..." The drawback to this cookie-cutter production is that overall health and vitality start to suffer, and recessive defects show up with alarming regularity.

Eventually there is nowhere to turn. Breeders refer to "having bred themselves into a corner." To regain fertility and hopefully rid their line of genetic defects they must now outcross (sometimes referred to as *linecrossing*). Pain and agony! We are not talking about using a different breed altogether but simply using an animal from another breeder's line. Oh, if it were that simple!

Most breeders try to at least choose an animal whose phenotype closely resembles their own line. Finding such a specimen is not easy when you consider that conformation, temperament, color, coat (or fleece) quality and other issues have to be carefully weighed and considered.

Novice breeders often naively think of the outcross as the answer to solving all their problems. They don't consider that the other line can carry genes coding for poor conformational traits and recessive defects as well. Australian German Shepherd Dog breeders introduced pituitary dwarfism into their kennels when they outcrossed to English and German lines after a forty-four-year-old import ban was lifted. Instead of hybrid vigor, outcrossing brought hybrid misery. The crossing of Irish Setters and Standard Poodles at a colony maintained by the University of North Carolina resulted in cases of subluxation of the carpus (wrist bones in the foreleg). Outcrossing is not always the sure path to health and happiness that novice breeders think it is. Breeders should also understand that when hybrids are bred back into the original line, any hybrid vigor accrued disappears rapidly.

The next step in such a program involves another controversial and much disputed breeding technique. Many breeders, after crossing out into a different line, will choose *backcrossing* (daughter bred back to sire or son back to dam) to retain the virtues of their original line. Breeders can hit the genetic jackpot with such a bold move, or plunge their program into such an abyss of problems that it will take years to recover.

It takes an enormous amount of time, study, knowledge of pedigrees and individual animals, as well as a portion of sheer luck to succeed in such a venture. Ruthless culling (total removal of all inferior stock from further breeding) must be practiced under such a program. Backcrossing after the initial outcross is, of course, not a necessity. To continue to outcross, though, will destroy the fixed phenotype of your line and all the years of work you've devoted to creating it. All breeding systems carry risks. There are no simple solutions.

How does this apply to our alpacas and specifically to the Suris? Creating a phenotypically uniform line takes years of breeding. Alpacas, especially Suris, have not been in our country long enough for many breeders to have created a line. Individual Suris are not even close to breeding true (not producing Huacaya offspring), let alone being part of a well-planned, well-bred, and well-performing line.

We can't speak of true outcrosses at this time. The Suri stud you're using for your "outcross" to a Huacaya might only be heterozygous for the Suri trait himself and possibly line-bred on the female without your knowledge. I am talking about imports here. Linebreeding does not have to be a deliberate decision on a breeder's part. It can and does occur in large herds allowed to breed at random. One such linebreeding however does not create a "line"!

Hybrid vigor, also called *heterosis*, is only created if a substantial difference in gene frequency exists between the parents. We're talking more than just a few generations! Alpacas with pedigrees the length of pencil stubs should not be used for such an ambitious project.

Listen to what Malcolm B. Willis (1989) has to say on the subject of heterosis: "When we are dealing with unrelated members of the same breed, it is unlikely that major genetic differences will occur between the individuals... They may look very different, but these differences will be for relatively minor issues in genetic terms, however crucial they may seem to be in the show ring...outbreeding within a breed is not therefore a means of making major advances in heterotic traits..."

When Dr. Willis wrote this about dogs, he was a Senior Lecturer in Animal Breeding and Genetics, at the University of Newcastle-upon-Tyne, Great Britain. Alpaca breeders might want to study his writings and draw their own conclusions.

Breeders of livestock often speak of *upgrading* their stock. Upgrading is a technique used by breeders who can't afford to start with top of the line breeding stock. They start out with animals of average or even poor quality and upgrade by breeding to superior studs.

The original stock may be of good or even excellent quality but not bred for the desired purpose. A sheep farmer might want to change from meat sheep to sheep with superior fleeces. He starts breeding his meat sheep to merinos, for example, and eventually his sheep will carry an "upgraded" fleece compared to the ones he started out with.

This is, to my way of thinking, a tortuous way to go about achieving a breeding goal. It also does not, in my opinion, apply to Huacaya-Suri crosses, unless one feels that Suris are inherently superior to Huacayas and should eventually be the only type. I personally find the existence of both varieties a wonderful genetic gift to be cherished and protected. I've read conflicting information on the value of the two different fleeces, so I will not comment on that aspect. Breeders should be aware though that such financially driven selection pressure (possibly resulting in the eventual disappearance of one variety) can have as yet unforeseen negative genetic repercussions.

Let's consider the relatively small Suri gene pool. When sheep breeders run into genetic trouble with their chosen breed, they have many different breeds to turn to for help. Suri breeders, unless they want to turn to llamas, can only use the Huacaya variety as a safety hatch. Why squander the potential genetic piggy bank held by the Huacayas in this country on something as frivolous (comparatively speaking) as color?

I personally find suggestions to remove Suri females out of Huacaya x Suri crosses from a breeding program a little harsh (after all, neither variety can be considered defective or undesirable). However, I must admire breeders who are willing to make substantial financial sacrifices for the benefit of reaching their goal — creating a line of Suris who will produce true to type.

Suri Genetics — Part II

The joy and happiness felt immediately after the birth of the first cria is short-lived. Incredulous, the breeder stares at his spanking new and now dry alpaca baby. Cushing next to her peacefully grazing Suri momma rests a healthy — Huacaya! With surprisingly swift and steady movements, the little fluff-ball rises and follows her dam across the expansive pasture. Gorgeous dark fawn fleece, beautiful conformation, four incisors fully erupted, straight and obviously strong limbs — what more could one ask for?

Still speechless, the weak-kneed breeder sits on a bale of straw, a thousand thoughts running through his head. This simply cannot be possible! When he and his wife had "discovered" alpacas, they had immediately fallen in love with the creatures who looked so much like big stuffed toys. As they continued to visit farms, the husband came to appreciate the unique appearance of the Suris he encountered. He had been assured that Huacayas and Suris were two separate breeds, with the Suri being the more rare — and therefore more expensive.

The argument for a more lucrative investment finally convinced both spouses. The higher purchase price seemed well worth it. Husband

and wife carefully chose two beautiful Suri females as their foundation animals.

The folks admired the highly advertised, obviously typey and well-bred Suri stud both of their girls had been bred to. Son of a much-coveted import, the stud had been purchased by his owner for what seemed to our novice breeders to be an astronomical sum of money. They had visions of their alpaca business being firmly under way, and being able to afford the purchase of such a costly animal. Perhaps one of their girls would give birth to a cria as spectacular as his sire!

And now — what were they to do with the little imposter born on their farm? And, for Pete's sake, how did it get there in the first place?

If readers expect a decisive answer from me, I must disappoint them. There is presently much speculation regarding the mode of inheritance of the Suri trait. I will present several of the most likely possibilities. It is then up to the individual breeder to arrive at conclusions based on their own observations and those of other breeders they trust to be knowledgeable in the field of genetics.

The uncertainties stem from the fact that many South American alpaca breeders either allow random field breeding or do not attempt to separate the two varieties in their breeding program. Blood-typing to verify parentage is also not employed in South America. Keeping pedigrees and verifying parentage under such circumstances is impossible.

Based on his own observations, Suri breeder David TenHulzen reports that "it appears that the Suri gene is dominant over the Huacaya gene." Well-respected authorities on breeding alpacas such as Dr. Raoul Ponzoni and Dr. Mick Carrick hold identical views. All three, however, reserve final judgement until further research has been conducted.

The inexperienced breeder might well ask how a comparatively small group of animals representing a different fleece type can carry the dominant gene. Wouldn't it seem to be the other way around?

Let's clear up this common misconception. *Dominance*, genetically speaking, does not always mean "more of." A specific gene coding for white, for example, is dominant in horses. Very few truly *dominant white* horses exist (not to be confused with grey horses who appear white as they age, or horses who are genetically black or red but <u>appear</u> white as the result of dilution genes).

Think of the dominant allele as either present or not. It cannot be hidden. If an animal doesn't express the dominant trait, it doesn't possess it in its genome and thus cannot pass it on. Dominant traits can easily be lost to a breeding program due to homozygosity resulting in embryonic death, environmental influences, and selection pressure used by breeders. It is child's play to select against them.

It also means that a Huacaya can't possibly produce a Suri unless bred to a Suri. A few breeders claim that is exactly what occurred on their farms, or that they have observed such crias in the pastures of other breeders.

It is, of course, genetically feasible for two Huacayas to claim legitimate parentage of a Suri offspring. Let's explore several potential models.

Mutations occur in all species. Dominant white, for example, is believed to occasionally crop up as a mutation in horses and then goes on to reproduce under the same genetic control as your straightforward, run-of-the-mill, easily predictable, autosomal dominant allele. Mutations resulting in marked changes of phenotype (as in the coat type of the Rex cat) are extremely rare in most species. Their role in the current surprise appearances of Suri crias is probably only a remote possibility — although the curly ringlets of the first Suri to ever be born in South America were more than likely the result of a mutated gene.

Other genetic models exist for our exploration. A somewhat complicated one is the example of black horses. Their phenotype is determined by the interaction of two separate loci (genetic "addresses") called, respectively, the *Agouti* and *Extension* locus. At the Agouti locus the dominant allele (**A**) produces a bay horse, the recessive (**a**) a black one. Now it gets a little trickier. At the Extension locus we find almost the opposite. The recessive allele (**e**) produces chestnut horses. Carried in its homozygous (**ee**) form, it masks the expression of black at the Agouti locus. This means the double dose suppresses the expression of black pigment, such as the "points" on a bay horse. The dominant allele (**E**), on the other hand, gives the Agouti locus alleles "permission" to fully express themselves. For a horse to be black it has to be **aa EE** or **aa Ee**. In other words, it has to be dominant at the Extension locus for the Agouti locus to "show its stuff."

I am not suggesting that inheritance of the Suri trait works in a similar fashion, nor should you assume that alpaca colors are inherited in exactly the same way. It does serve as an excellent illustration of how genes from different loci sometimes must work in tandem to produce desired results.

Is a modified version of this inheritance mode a possibility in the case of our Suri trait? Perhaps.

Sometimes two separate loci code for the identical phenotype. In Dr. Sponenberg's book *Equine Color Genetics* (1996), I learned of the existence of such a trait. At least two independent genes can lead to curly horses. One of them is due to a recessive mutation and produces curly coats in normally straight-haired breeds (Missouri Fox Trotters and Percherons). The other gene resides at an entirely separate locus and is under dominant genetic control.

Sponenberg goes on to tell us that horses carrying the dominant allele are occasionally only minimally curly. It is thus easy for casual observers to miss the trait. A fully curly foal out of such a horse might then come as a surprise to unsuspecting and unobservant breeders.

We can further speculate that the Suri trait is *incompletely dominant*. This concept is easy to understand because nothing is hidden. Each genotype has its own phenotype. I'll use color inheritance to clarify this concept. In dogs, for example, the **B** locus carries two alleles. **B** allows black pigment to form, and **b** modifies it to liver color. If the Agouti locus permits the expression of black pigment, **BB** and **Bb** dogs will be black, **bb** will be liver. Such *complete dominance* is "normal" for a dominant gene.

As an example of *incomplete dominance*, the **C** locus also gives us two alleles (**C** and **c**), resulting in three genotypes as well: **CC**, **Cc**, and **cc**. The difference here is, as stated earlier, that each genotype has its own phenotype. **CC** allows full expression of color. **Cc** dilutes the base color red to fawn, for example, while **cc** results in severe dilution, often to the point of making the animal appear white. This is a somewhat simplistic presentation as the **C** locus in dogs in reality has more than two dilution alleles (although the individual animal only carries two, of course), but it illustrates the concept very well.

Any incompletely dominant genetic pattern makes life easy for a breeder because heterozygous (as in **Cc**) animals can be recognized as such. The two alpaca fleece varieties (not breeds!) are very distinct. Some alpaca breeders, however, have reported owning or seeing alpacas with a sort of "in between" fleece.

To round out our smorgasbord of genetic options, I might as well mention another mode of inheritance — *incomplete penetrance.* Here we have the three possible genotypes again: **BB**, **Bb,** and **bb**, but **Bb** occasionally does not reveal its dominance. Phenotypically speaking, the animal looks like **bb**, but goes on to produce as **Bb**. In *Genetics of the Dog,* Malcolm B. Willis cautions us to be careful about the assumption that incomplete penetrance as it pertains to one gene applies in all such cases. He feels that *polygenic inheritance* may play a role instead, meaning that a cluster of genes must work together to produce the trait. Many conformation traits are indeed inherited under polygenic control. All traits governed by polygenes are subject to various degrees of penetrance.

At this time, I won't even venture a guess which of these possibilities might apply. Suri breeders, especially those with large herds, are better equipped to arrive at conclusions based on their observations. Let me assure you that the genetic patterns of inheritance do exist. It is not a matter of "hit or miss."

An article printed in a national publication, as well as various messages posted on the Internet's Alpaca Site, have given the impression that "cross-breeding" Suris and Huacayas will lead to rapid progress towards turning your Huacaya herd into a more lucrative herd of Suris. For the sake of our discussion here, let's adopt the theory that the Suri trait is inherited under an autosomal dominant mechanism. We have learned that a dominant trait is easily bred out of your stock. Unfortunately in our case, it's the recessive (Huacaya) allele we're trying to lose, which is a much more complicated proposition. Recessive alleles can be carried over many generations without being expressed.

Often breeders labor under the misconception that a heterozygous (**Ss**) animal always passes on the dominant trait. Not true! When an alpaca is **Ss**, only the **S** is expressed (unless the case of incomplete penetrance applies), but both **S** or **s** have an equal chance of being passed on to the offspring.

On the Internet site, a Suri breeder mentioned that over 90 percent of the herd would be homozygous for the Suri trait after only four generations of such a "crossbreeding" program. No breeder can or should give such assurances at present. His statement hinges on the presumption that the Suri sires used in the program will be homozygous (**SS**). We have no such guarantees.

Let's say you enjoy the blind luck of choosing such a genetic jewel. In the first generation all the crosses will result in phenotypically correct Suris, according to Mendelian principles. Your luck might run out in the second generation. The next sire you use to continue your program could just as well be heterozygous, despite all assurances to the contrary by his either not very knowledgeable (or plainly dishonest) owner.

We now have the following situation, with the Punnett Square demonstrating the probabilities of possible genotypes and phenotypes:

		sire	
		S	s
dam	S	SS	Ss
	s	Ss	ss

If it's your lucky day, the cria will be **SS**. If it's **Ss**, you have a phenotypically correct Suri but have made no genetic gain from the last gene-ration. If Murphy's Law prevails (what *can* go wrong *will* go wrong), your Suri mom gives birth to a Huacaya (**ss**). You are now exactly where you started — several years ago!

Somewhat accurate predictions about the possible percentages of heterozygous animals and those homozygous for the recessive trait can be made according to the formula called the *Hardy-Weinberg Law*. However, this only applies to randomly bred populations. Any time breeders have already applied selection pressure for certain traits, the formula doesn't work.

The *Theory of Mathematical Probabilities* certainly applies here (see Chapter Eight).

I decided to skip any detailed examination of possible financial and legal repercussions of such a proposed "crossbreeding" scheme. The majority of alpaca breeders seem to be business wise and financially astute people. They don't need me to point out the obvious.

People new to breeding animals often have difficulties with the uncertainties and all the unknowns of genetic science. Many are accustomed to controlling their own destiny through hard work and dedication.

Enter alpacas, llamas, horses, dogs, rabbits — whatever! Animals are living, breathing chemical factories. Unlike the owners of a factory manufacturing cars or toys, we simply don't have complete control over the "creative" process and cannot regulate everything to go our way. We can, however, educate ourselves and deal with genetic probabilities to the best of our ability. At present, it might be most practical to treat the Suri trait as possessing complete dominance and to make breeding decisions accordingly.

As for the shaken and weak-kneed breeder we left sitting on the straw bale — someone hand that man a cup of coffee (or something stronger perhaps?), and tell him to relax. There is much to be thankful for when we look at that healthy little cria.

Addendum to Suri I and II:

"Here, the information you requested from the ARI just arrived." I eagerly picked up the E-mail message my husband placed on my desk. I had contacted the Registry to find out exactly how many Suris had been registered out of Huacaya parents.

In correspondence dated December 6, 1999, Alan Schmautz wrote: "Dar Wassink has passed along your request concerning Huacaya parents that have Suri offspring. There are 8 documented cases in the ARI database." Under "subject," Alan had appropriately listed: *Huacaya+Huacaya =Suri*.

Eight babies! How many Suris were registered altogether? Several months earlier, I requested this information for an article I wrote for the German *LAMAS* publication.

At that time, Dar reported approximately 3,000 Suris in the Alpaca Registry records. As we enter the year 2000, Suri registrations have increased to slightly more than 4,000. Mathematically, we can express the 8 to 4,000 ratio as 0.2 percent (two-tenths of one percent).

Should these eight crias influence how we view the inheritance mechanism of the Suri fiber? I don't think so. If anything, the rarity of their appearance supports the opinions of those breeders and respected authorities who believe the trait is under autosomal dominant genetic control.

Exceptions to the rule are part and parcel of genetic reality. Novices should not allow such exceptions to cloud their thinking and prevent the practical application of a workable genetic model. In other words, don't become so bogged down in theory and so fretful about occasional deviations from the norm that you lose sight of the bigger picture.

Abnormalities of the **X** and **Y** chromosomes occur in roughly one out of 500 human beings. Do we raise our eyebrows when scientists refer to females as **XX** and males as **XY**? No, of course not. The fact that many people carry unusual combinations of these two chromosomes does not diminish the validity of the above-mentioned facts as they pertain to the norm.

One alpaca owner, informed of my research, wrote to me, "In light of those registrations, the inheritance of the Suri fiber doesn't seem all that simple." Gee, partner, eight babies out of 4,000 is a complication? You could have fooled me!

Their very existence will certainly keep us on our toes. It is proof that Mother Nature, however tightly harnessed by humans, is still Mistress of her domain. If you absolutely can't tolerate any uncertainties whatsoever in your life, breeding animals will surely send your blood pressure soaring.

In *Suri II*, by the way, I reported the claim made by several North American breeders that they have observed animals with "in between" fleeces. Jane C. Wheeler, Ph.D., also mentions the existence of such *chili*, telling us about animals with "intermediate wool characteristics," although she says that they are only seen "occasionally." Her article "Evolution and Origin of the Domestic Camelids" (*The Alpaca Registry Journal*, Vol. III, No. 1) offers valuable information to camelid breeders.

Rigoberto Calle Escobar writes in *Animal Breeding and Production of America Camelids* that "...no notoriously intermediate progeny are produced..." and "...on the contrary, as a result of indiscriminate breedings, offspring that phenotypically respond to perfectly defined types of either Suri or Huacaya are produced." He quotes prominent alpaca breeder Julio Barreda as calling it "a real miracle" and relates how Barreda marvels at the fact that the "ethnic features are still presented for both varieties, when from remote epochs and even to date, both have been bred mixed and together."

At present, I do not advise buying two Huacayas in the hopes of producing a Suri cria. You could wait forever, unless your luck factor is like the one little Charlie Bucket enjoys in *Charlie and the Chocolate Factory* (a story by Roald Dahl). Maybe you'll even hear the Oompa-Loompas singing in your barn.

Postscript — June 2003

It occurred to me that some readers may misinterpret what I have written. I do not consider the birth of a Huacaya cria out of two Suri parents a "problem" or a calamity. After all, Huacaya fiber is not a defect, and the cria is as "pure" a Huacaya as any other Huacaya alpaca. Would I purchase one if the right opportunity came along? Absolutely!

I do feel strongly that breeders must be honest with buyers purchasing their stock and present such a birth as a distinct possibility in any Suri breeding program. There is, in my opinion, nothing "dishonest" about a Huacaya-Suri crossbreeding program — as long as breeders offer full disclosure and don't lure potential buyers with false guarantees of quick success.

Since I wrote the original articles, Dr. Sponenberg has done extensive research into the inheritance of the two fiber varieties. I addressed his findings in Chapter 12. They certainly shed a new light on the issue. What about my earlier statement that a dominant allele cannot be hidden? While that is generally true, it certainly would no longer apply to the expression of the dominant Suri allele. If Dr. Sponenberg's hypothesis is correct, the suppressor allele does most definitely "hide" the effect of the dominant Suri allele. There are, as I've pointed out numerous times, always exceptions to genetic "rules."

My rather flippant comment to the reader who considered the few Suri babies out of two Huacayas a complication landed me in hot water (or dung pile) as well. Yes, the inheritance of Suri fiber is obviously more complicated than the Suri-Huacaya choice at one locus. However, from a <u>practical</u> stand-point, breeders need not fret over these "complications." Choices and breeding results boil down to:

> 1. Suri bred to Suri = you will get mostly phenotypical Suris but plan on getting some Huacayas. Dr. Sponenberg's data shows a 14 percent Huacaya rate (1980 total).

> 2. Suri bred to Huacaya = you will get Suris, but (speaking in probabilities) your chances of getting Huacayas are far higher. The data listed by Dr. Sponenberg gave a ratio of 89 Huacayas/56 Suris for 145 Huacaya/Suri matings — quite different from the "get rich quick" results promised by one author who urged breeders to cross Suris with Huacayas for financial gain.

> 3. Huacaya bred to Huacaya = Huacayas. If you get a Suri out of this combination, immediately start playing the lottery. You beat the odds!

All statistics aside, you can't go wrong by thinking of the Suri trait as dominant. Now that isn't complicated, is it?

Chapter Thirty-five

ADVICE (SHORT AND TO THE POINT) AND APPLE CAKE

CHAPTER THIRTY-FIVE

Advice (Short and to the Point) and Apple Cake

You can read volumes, study pedigrees and animals, and question other breeders about their beliefs. In the end you stand alone. You must make the choices and decisions, set the priorities, and prepare to enjoy happy outcomes or suffer any negative consequences of your actions.

You must decide whether inbreeding, linebreeding, or outbreeding is the best road to take. The best advice I can give you is to immerse yourself in studying all aspects of your chosen species and/or specific breed.

"But," some of you might complain, "the book never mentioned that … and you forgot to tell the reader…in Chapter 10 you should have added…" Hold on to your hat — or two hats for that matter! This book is meant to be an <u>introduction</u> to genetics and give breeders a <u>basic</u> working knowledge of very complex subject matter.

To those of you who have discovered a completely new world to explore, we can say that the subject of genetics is much more complex than presented in this modest guide. We have simply led you, as promised, through the first steps of your journey into a science as it relates to breeding mammals. There remains a vast store of knowledge for you to discover. Venture out and find it!

Above all, enjoy it — your newfound knowledge, the camaraderie of fellow breeders, the preparations leading up to a birth, and finally, the new life you helped to create.

Talking about camaraderie: you will find that many breeders are very social people. They like to get together with others who share their hobby — well, let's call it a passion. Food that helps to keep body and soul together is an important component of such gatherings. Here is the recipe for a tart German apple cake you can serve when you host your local club meeting.

Ingrid Wood and Denise Como

Stormwind's Apple Cake

<u>Dough</u>
2⅔ cups unbleached flour
¾ cup sugar
1 pinch salt
2 eggs
3½ sticks cold butter

<u>Filling</u>
1 cup raisins
9-10 large Granny Smith apples
juice of 1 lemon
1 cup apple juice
2 teaspoons cinnamon
1½ packages vanilla pudding (<u>not</u> instant)
3 cups half & half (cream)
½ cup sugar
1 tablespoon vanilla extract

<u>Dough</u>
1. Mix flour, sugar, and salt. Add eggs and butter (cut butter in small pieces). Knead until dough is smooth and all ingredients are blended.
2. Form a ball with the dough, wrap it, and put it in the fridge for 30 minutes.
3. Divide dough into two portions.
4. Butter a casserole dish.
5. Roll out one portion of the dough and spread it evenly in the casserole dish.
6. Preheat oven to 375 (F) degrees.

<u>Filling</u>
1. Cut apples in half and remove cores. Simmer apples, apple juice, and raisins. Apples should remain fairly firm. Drain juice. Pour lemon juice over apples. Cover rolled out dough with apple/raisin mixture.
2. Cook pudding (using the 3 cups of half & half). Add vanilla extract. Pour mixture over apples.

<u>Finish</u>
Roll out second portion of dough, cover filling and seal edges. Cut vent holes in the top. Bake cake for 60-75 minutes. Brush egg yolk over top about 15 minutes before cake is finished.

Cover and keep in a cool place, but do not refrigerate. Serve the next day with whipped cream.

Do you think a cake recipe in a book about genetics is crazy? You're right, it is! Gregor Mendel, who appreciated good food, would have loved this cake. Enjoy it with your breeder friends or customers who will be visiting to pick out that special puppy, arrange for a stud service to your sensational ram, or to see the brand new llama cria in your pasture. Socializing with other breeders is part of the fun.

Denise and I hope that we've opened your mind to new possibilities and wish you well in all your breeding endeavors!

Postscript

A BREEDER'S STORYQUILT

POSTSCRIPT

A Breeder's Storyquilt

Novice breeders, like growing children, pass through many stages. Only after years of extensive study are breeding objectives and long-range goals clearly etched in their minds.

These passages are inevitable, beneficial, full of lurking dangers, exciting, exasperating, enlightening, demoralizing, uplifting — in short, they are a microcosm of the growth we experience in real life. It is not, in my opinion, desirable or advisable to skip a stage. They all serve an important purpose in our quest to breed superior stock.

The first stage is the easiest one. Oh, the naive innocence of those early days! As a mere babe-in-arms you are blissfully "asleep" most of the time. During brief periods of alertness, you stare in wonder and delight at all the animals. You recognize broad differences, but far be it from you to pass judgement. Your admiration is boundless. Those handsome adults, those precious babies — how is one to choose? "I can't decide. Pick for me. They're all so perfect." The old breeder smiles. "No, they're not. Let me show you." She explains, points, demonstrates, and lectures. You blink in confusion, and your attention wanders. Isn't the one sitting down so cute? And look at those two young ones over there wrestling and playing! Who ever saw anything more entertaining? Don't you just love them all? And when is that old breeder gonna stop picking on her own animals? She doesn't even look close to a perfect "10" herself, for crying out loud!

Of course, you must admit, she does seem to know what she's talking about. You decide to take the plunge. Puffed up with pride and giddy with excitement you eventually start life as an owner. You're thrilled with your choice, waking up every morning grateful to a fate that bestowed such wonderful creatures on you. Come to think of it, they *are* "just like potato chips" — you can't have just one or two!

Slowly your little group expands. The old breeder continues to call and support you. Together you decide on a stud for your first breeding. Routine maintenance and training chores have become, well, routine. Your confidence soars, and you feel absolutely ready to take on added responsibilities. Occasionally you question a decision or choice the old breeder makes. With somewhat unsteady toddler steps, you venture out into the vast world of health and genetic issues — listening, learning, memorizing.

Though you feel the need to assert your independence, you're glad to return to the secure presence of one who has listened longer, learned more, and memorized a vast array of details. The first babies arrive. "Come and see them! They're just gorgeous! I wouldn't change anything about them!"

The old breeder visits and studies, watches and probes, inspects and selects. She doesn't agree with you. There are certain things she would change if she could, and she tells you about them with her usual frankness. Well, you hate to admit it, but you must confess that your old friend is right. "Don't look down in the dumps," she admonishes, "you have much to be happy about." Yes, it's true. Before your mentor pointed out faults, she had extended sincere praise, admiring the fact that your chosen stud had improved on certain traits in the original stock purchased from her.

But it's too late. Your carefree years of "childhood" are over. You've entered the ugly teenage phase of your development as a breeder. Your eyes are now fully open, and some days you do not like what you see. Lord have mercy, how could you not notice how cow-hocked that female is? And one of the babies you thought was so stunning — look at that ugly head, you might as well cover it with a bag. You still love your animals, the pathetic little creatures. If you don't, who else will? Look at that one — with ears only a mother could love! Over there walks the female you had decided was so beautiful you would never sell her. Doesn't it look like she's paddling in the rear now? Why has the whole business of breeding all of a sudden become so complicated? Confusion reigns.

Time passes. Your females, who looked so ugly yesterday, don't seem so bad today, maybe even downright pretty. But now a new problem has cropped up to vex and aggravate you — are there any decent studs "out there" deserving of your beautiful girls? Not a one that you can find! This one has no bone, that one's bite is overshot, the third one gaits as if his rear legs are tied together. What's wrong with their owners — they have the nerve to ask for a stud fee? *They* should be paying *you* for the honor of even considering the use of their males on your lovely females!

The old breeder cautions you not to become a "fault finder." You don't understand. Wasn't she always pointing out in minute detail what was wrong with her own stock? She chooses her words with care. "I am very critical of my animals, as you should be of yours, but if you notice, I look at the total animal and its superior qualities first. I register and evaluate faults in the 'second round' and try to put them in proper perspective."

You stare thoughtfully at your animals and remember snippets of past conversations. "Don't show under that judge, he's a tooth fairy. He'll pass over the most spectacular and correct specimen with one missing tooth and put up a mediocre one because the latter has full dentition." "It's a crying shame no one ever bred to that male. He was a little straight in the front, but did you ever look at the rest of him? He had so much to offer."

You don't know it yet, but you're about to leave the *"Angst und Weltschmerz"* phase behind you and enter the last passage. You've paid your dues. Countless hours of studying pedigrees and scrutinizing live animals has cleared your mind. You have a crystal-clear mental image of the perfect specimen. You realize it's a goal never to be achieved, but you will continue to travel the long road in its pursuit.

Ups and downs, lumps and bumps — now you know they are a vital part of every breeding program. The old breeder heard the loud thump when you fell off the High Horse of novice arrogance, but was too tactful to comment. What a fool you were — passing simplistic judgements and commenting on breeding decisions more experienced breeders made before you understood the "whole picture" in all its complexity. Never did you dream of the amount of work and dedication involved in becoming a true breeder, not merely a merchant of animals. Never did you dream you would enjoy it so much and thrive on the challenges.

Prospective clients come to visit. They clap their hands in delight. "What beautiful, beautiful animals! They're all so perfect." You smile. "Thank you, but they're not really perfect. Let me explain…"

Glossary

Allele - an alternative form of a gene.
Autosome - any chromosome other than the sex chromosome.
Chromosome - threadlike bodies in the nucleus of a cell that contain the genes.
Co-efficient of Inbreeding - the result of a mathematical formula calculating the degree of inbreeding.
Cross-eyed & Crook-fingered - Denise, after deciphering & typing Ingrid's handwritten manuscript and corrections. (Denise: *"Amen to that!"*)
Crossbreeding - the mating of animals of different breeds.
DNA - Deoxyribonucleic Acid.
Dominant - the allele masking the presence of others at the same locus.
Epistasis - the situation whereby a gene at one locus can influence the expression of genes at other loci.
Gamete - egg or sperm cell of a parent.
Gene - unit of inheritance that passes traits from parents to offspring.
Genetics - the science of heredity.
Genome - all the genetic material in an individual, breed, or species.
Genotype - the genetic make-up of an individual, not necessarily all visible.
Germ Cells - egg or sperm cells.
Hemizygous - a gene present in a single dose such as in a haploid cell.
Heterosis - the performance boost resulting from the crossing of two breeds or totally unrelated lines.
Heterozygous - different, such as **Aa** or **Bb**.
Homozygous - the same, such as **AA** or **bb**.
Hybrid Vigor - see heterosis.
Hypostasis - the situation whereby a gene at one locus is influenced by the expression of genes at other loci.
Inbreeding - the mating of related animals where neither individual is the ancestor of the other.
Incomplete Dominance - each genotype expressing its own phenotype.
Incomplete Penetrance - a heterozygous gene such as **Aa** expressing itself as **aa** but reproducing as **Aa**.
Linebreeding - the type of breeding which concentrates on one given ancestor.
Locus (plural: **loci**) - a specific location on a chromosome.
Meiosis - germ cell formation, number of chromosomes is reduced to half.
Mitosis - ordinary cell division, does not reduce number of chromosomes.

Mutation - process by which a gene undergoes a structural change.

Outbreeding - a mating between individuals that are less closely related than average.

Phenotype - genotype and environment combine to form the phenotype (appearance).

Polygenic - character controlled by two or more genes.

Punnett Square - a diagrammatic way of presenting the results of a random fertilization.

Recessive - a recessive allele needs to be inherited in duplicate to express the trait.

RNA - Ribonucleic Acid.

Sex Chromosome - the chromosomes determining the sex of the offspring; choices are **XX** (female) or **XY** (male).

Somatic Cells - cells that become differentiated into the tissues, organs, etc. of the body.

Transposons - a gene that can move around the genome, moving either within or between chromosomes.

Wood's Law - do not inbreed or linebreed if you have difficulty comprehending the formula for the co-efficient of inbreeding.

Zygote - the first living cell formed by the union of two gametes (fertilized egg).

References

The Complete Alpaca Book (2003)
Eric Hoffman & Selected Contributors
Bonny Doon Press, publisher
121 McGivern Way, Santa Cruz CA 95060
email: bd1alpaca@aol.com

Alpacas Magazine
PO Box 1992, 1140 Manford Ave, Estes Park CO 80517-1982

American Kennel Club Gazette (monthly magazine)
5580 Centerview Drive, Raleigh, NC 27606-3390

Animal Breeding and Production of American Camelids (1984)
Rigoberto Calle Escobar, author
Ron Hennig-Patience, publisher
(to order, phone 1-800-245-2627)

Bailliére's Comprehensive Veterinary Dictionary (1988)
D.C. Blood & Virginia P. Studdert, authors
Baillière Tindall, publishers (UK)

A Conservation Breeding Handbook (1995)
D. Phillip Sponenberg & Carolyn J. Christman, authors
The American Livestock Breeds Conservancy

Discovery, The Search for DNA's Secrets (1981)
Mahlon B. Hoagland, author
Houghton Mifflin Company, publisher

Equine Color Genetics (1996)
D. Phillip Sponenberg, D.V.M., Ph.D, author
Iowa State University Press, publisher

GALA Newsletter, The (Greater Appalachian Llama and Alpaca Assn)
Peru Farm - 661 High Street, Athol NY 12810
(518) 623-3987

Genetics and the Behavior of Domestic Animals (1998)
Temple Grandin, editor
Academic Press, publisher

Genetics for Cat Breeders (1971, 1973)
Roy Robinson, author
Pergamon Press, publisher

Genetics of the Dog (1989)
Malcolm B. Willis, author
Howell Book House, publisher

Genetics — The Mystery & The Promise (1992)
Frances Leone, author
TAB Books, McGraw Hill Inc, publisher

How to Breed Dogs (1937, 1947)
Leon F. Whitney, author
Howell Book House, publisher

Irish Wolfhound Guide, The (1973)
Alfred W. DeQuoy, author
Cahill & Company, Ireland, publisher

LAMA LETTER, The (PA Llama and Alpaca Assn)
T. Audean Duespohl, editor
RD #1, Box 418, Seneca PA 16346

Llama Life II
5232 Blenheim Rd, Charlottesville VA 22902
(804) 286-2288

Language of Genes, The (1993)
Steve Jones, author
Anchor Books, Doubleday, publisher

Medicine and Surgery of South American Camelids (1998)
Murray E. Fowler, D.V.M., author
Iowa State University Press
to order, phone 1-9-800-862-6657

The Merck Veterinary Manual (1998)
National Publishing, Inc.
(to order, phone 1-800-635-5262)

Nature of Life, The (1989)
John H. Postlethwait & Janet L. Hopson, authors
McGraw Publishing Company

Neuweltkameliden (1997) (German)
Dr. Matthias Gauly, author
Parey Buchverlag, Berlin

Raising Sheep the Modern Way (1989)
Paula Simmons, author
Storey/Garden Way Publishing

sheep! Magazine
PO Box 10, Lake Mills WI 53551
(800) 272-4628

Second Creation, The (2000)
Ian Wilmut, Keith Campbell, Colin Tudge, authors
Farrar, Straus & Giroux, publisher

Sighthounds Afield — The Complete Guide to Sighthound Breeds and Amateur Performance Events (2003)
Denise Como, author
1st Books Library, Bloomington IN
order from: www.1stbookslibrary.com

Running Your Sighthound (6th revision of the handbook **So, You Want to Run Your Sighthound**, 2003)
Denise Como, author
e-mail: wolfwindbz@frontiernet.net

VetGen
1-800-483-8436, e-mail: HealthyDog@VetGen.com

Windhunde (1979) (German sighthound book)
Ingeborg & Eckhard Schritt, authors
Franckh'sche Verlagshandlung, publisher

> Genetics Workshops for Breeders of Mammals:
> (Groups or private instruction)
> **Stormwind Alpacas**
> Ingrid Wood
> 1862 Jacksonville-Jobstown Rd, Columbus, NJ 08022
> (609) 261-0696 - alpacas@uscom.com — www.StormwindAlpacas.com

Ingrid's Letter to Our Readers

Dear Readers,

Writing this slightly (?) unconventional book was educational for me as well as a pleasure. To my amazement and immense relief, Denise was able to decipher the original, very untidy, handwritten manuscript as well as survive numerous additions and changes. Her corrections, contributions, and advice throughout the long process were always pertinent and to the point, yet reassuring and supportive at the same time. Although most of the research and writing is mine, Denise fully deserves to be listed as contributing co-author. (My husband offered to type the manuscript. I decided thirty years of marriage should not be sacrificed to a book. I've always had a firm, practical grip on priorities in my life.)

Once again, I thank the manuscript reviewers who furnished me with many constructive suggestions for improvement. Denise and I also welcome comments — both positive and negative — from our readers.

To tell you the truth, I am plumb tuckered out from the intense concentration this project required. I am ready to try something lighter, something less intellectually taxing — quite possibly a historical novel with a breeder as the hero or heroine. Maybe I'll just sit in the yard, play with my Whippets, and watch the alpacas graze and frolic.

I found this quote in a book on ranching:

> "I think I could turn and live with animals, they are so placid and self-contained …they do not sweat and whine about their condition."
>
> Walt Whitman

Thank you for reading our book — writing it has been an adventure.

Ingrid Wood
Springfield, New Jersey, 2003

Index

achondroplastic, 161
adaptations, 163
additive gene action, 83, 84, 85
Afghan hound, 36
agalactia, 37, 152
Agouti, 12, 199, 210, 211, 212, 213, 224, 225, 226, 227, 228, 230, 231, 246, 248, 256, 282, 283
Agouti locus, 210, 211, 212, 224, 225, 228, 230, 231, 246, 248, 256, 282, 283
AKC, x, xiii, 74, 109, 152, 156, 169, 172, 174, 175, 177, 191, 229, 255, 256, 265, 271, 313
AKC Gazette, x, xiii, 74, 109, 152, 156, 169, 174, 175, 177, 191, 229, 256, 271, 313
albino, 13, 105, 209, 218, 245, 247, 250, 263
allele, 7, 20, 21, 22, 29, 33, 41, 42, 49, 50, 51, 53, 58, 59, 60, 63, 67, 71, 72, 77, 79, 80, 81, 86, 103, 140, 149, 150, 162, 198, 205, 209, 210, 211, 212, 213, 214, 217, 218, 224, 226, 227, 228, 229, 230, 231, 233, 234, 237, 238, 240, 243, 245, 247, 253, 254, 255, 260, 261, 282, 283, 284, 288, 299, 300
alpaca, v, x, 3, 5, 8, 13, 17, 22, 38, 60, 72, 74, 78, 81, 94, 100, 109, 117, 118, 124, 133, 139, 151, 158, 161, 168, 171, 172, 174, 176, 181, 190, 191, 192, 194, 197, 198, 199, 201, 206, 220, 222, 223, 225, 226, 227, 228, 229, 230, 232, 233, 234, 236, 237, 238, 239, 242, 243, 245, 246, 247, 248, 250, 259, 260, 262, 263, 264, 265, 266, 275, 276, 280, 281, 283, 284, 286, 287, 288
Alpacas Magazine, 25, 27, 301, 313
American Kennel Club, 301
Anderson, David E., 262, 263
appaloosa, 240, 243, 245
ARI, 73, 79, 221, 222, 224, 230, 235, 236, 237, 238, 240, 241, 286
ARI Journal, 222, 224, 240, 241
assortative, 97
atresia ani, 151
autosomal, 33, 41, 42, 64, 74, 78, 83, 168, 216, 260, 263, 282, 284, 287
autosomes, 33, 93
Avery, Oswald, 26
BAER, 261, 262
Bell, Jerold S., 85, 145
Beyer, Nina R., vi, 89, 90, 203, 204, 215
Blatchford, Lynd, 176
Borzoi, vi, 3, 6, 7, 15, 85, 100, 145, 167, 177, 184, 193, 197, 199, 203, 255, 271, 313
Bowling, Sue Ann, 211, 213, 256
breeds, xii, xiv, 3, 4, 5, 7, 22, 36, 49, 60, 74, 85, 98, 116, 129, 131, 132, 133, 144, 157, 162, 168, 172, 175, 181, 183, 185, 194, 199, 204, 210, 211, 213, 216, 218, 227, 249, 265, 275, 276, 277, 280, 283, 284, 299, 313

Bruford, Michael, 5
Caesarean, 172
camelid, vi, x, 5, 17, 78, 85, 95, 98, 152, 177, 191, 194, 198, 206, 220, 223, 224, 225, 228, 231, 232, 234, 236, 238, 242, 243, 247, 249, 265, 275, 288, 313
Campbell, Keith, 6, 149, 174, 303
carriers, 13, 78, 133, 149, 150, 153, 156, 157, 158, 168
cats, 7, 12, 36, 49, 174, 181, 219, 220, 227, 249, 260, 263
cell, xii, 15, 16, 17, 18, 19, 20, 26, 27, 28, 29, 33, 34, 38, 42, 44, 45, 94, 107, 135, 138, 155, 156, 165, 261, 299, 300
Chase, Martha, 26
choanal atresia, 109, 151
chromosomes, 17, 18, 20, 21, 22, 23, 26, 33, 34, 35, 77, 91, 92, 93, 138, 287, 299, 300
cochleosaccular degeneration, 260, 261, 262
co-dominance, 68
co-efficient of inbreeding, 11, 104, 118, 121, 124, 127, 128, 300
color inheritance, v, xiv, 7, 12, 22, 60, 83, 197, 198, 201, 204, 205, 206, 210, 216, 217, 220, 233, 234, 237, 247, 249, 261, 283
Craven, Pat, v, 25, 27
cremello, 67, 229, 259, 260, 262, 263, 264, 265, 266
cria, 17, 36, 42, 78, 98, 117, 118, 151, 183, 190, 197, 222, 223, 227, 236, 239, 246, 247, 248, 264, 276, 280, 281, 285, 286, 288, 293

Crick, Francis, 26, 36
crossbreeding, 3, 131, 133, 145, 175, 276, 277, 285, 286, 288
cryptorchid, 151, 269
cryptorchidism, 151, 269
culling, 106, 127, 277, 278
Cystic Fibrosis, 6, 13, 149, 150
DeQuoy, Alfred W., 103, 104, 116, 127, 302
dilution, 67, 202, 209, 214, 215, 218, 229, 230, 231, 245, 253, 260, 265, 282, 283
diploid, 18, 19
disassortative, 97
DNA, iii, vi, xii, xiii, 5, 16, 17, 18, 19, 20, 23, 24, 25, 26, 27, 28, 29, 33, 34, 35, 37, 44, 77, 78, 94, 103, 106, 109, 118, 151, 152, 154, 155, 156, 157, 158, 161, 162, 163, 164, 168, 175, 221, 223, 228, 239, 299, 301
DNA testing, 77, 109, 156, 157, 175, 221
dogs, vi, xiii, 5, 33, 49, 60, 64, 98, 99, 109, 114, 115, 134, 136, 137, 158, 162, 167, 169, 171, 172, 178, 182, 183, 184, 191, 192, 194, 198, 204, 210, 212, 213, 214, 215, 216, 217, 219, 224, 227, 228, 232, 237, 240, 253, 254, 255, 262, 265, 270, 271, 275, 276, 279, 283, 286, 313
Dolly, 6, 19, 45
dominant, 6, 7, 29, 33, 42, 49, 50, 51, 52, 53, 58, 59, 60, 63, 64, 67, 68, 71, 72, 74, 78, 79, 80, 81, 83, 85, 149, 150, 162, 198, 199, 201, 202, 205, 210, 211, 214, 215, 216, 218, 219, 222,

223, 224, 225, 226, 227, 228,
  233, 234, 236, 237, 241, 243,
  244, 245, 246, 247, 248, 253,
  256, 257, 259, 260, 261, 281,
  282, 283, 284, 287, 288, 289
donkey, 3, 132, 205
Drosophila, xiii, 37
enzyme, 6, 27, 162, 260
epistatic, 205, 212, 217, 255
Equine Color Genetics, 3, 205,
  226, 227, 228, 237, 238, 244,
  250, 263, 283, 301
Escobar, Rigoberto Calle, 4, 18,
  78, 79, 174, 176, 288, 301
eumelanin, 201
Evans, Norm, 194
Ewing, Barbara, ii, vi, 50, 137,
  197, 203, 237, 255, 256, 257
Extension locus, 212, 213, 226,
  227, 255, 282
extreme white piebald, 203, 217,
  222, 223, 240, 245, 246
fertility, 84, 143, 144, 194, 270,
  277
Finn sheep, 36
freemartins, 36
Gala Newsletter, 272
gametes, 17, 22, 300
gene, xiii, 6, 7, 17, 20, 25, 27, 28,
  33, 36, 37, 41, 42, 43, 44, 45,
  49, 50, 52, 59, 63, 64, 67, 71,
  73, 74, 79, 83, 85, 86, 87, 93,
  103, 127, 133, 135, 136, 145,
  146, 150, 151, 155, 156, 157,
  158, 161, 164, 181, 184, 190,
  198, 199, 200, 201, 203, 205,
  209, 210, 214, 215, 216, 226,
  227, 228, 231, 234, 238, 239,
  241, 244, 245, 247, 260, 261,
  262, 263, 266, 279, 280, 281,
  282, 283, 284, 285, 299, 300

genome, xiii, 11, 18, 19, 21, 22,
  30, 41, 42, 53, 85, 93, 108,
  139, 140, 150, 158, 163, 190,
  272, 282, 300
genomic imprinting, v, xii, 41, 42,
  43, 45, 46
genotype, 8, 20, 22, 50, 58, 67,
  71, 79, 106, 110, 115, 116,
  132, 201, 215, 216, 245, 246,
  256, 271, 283, 299, 300
Gerken, Martina, 5
German Shepherd Dogs, 11, 85,
  211
Graham, Dale, vi, 198, 201, 202,
  204, 224, 226, 227, 230, 236,
  237, 240, 243, 244, 245
Grandin, Temple, 86, 190, 203,
  267, 302
grey, 86, 164, 197, 198, 199, 200,
  211, 214, 215, 223, 227, 229,
  230, 231, 232, 233, 234, 235,
  236, 237, 238, 239, 245, 246,
  248, 254, 265, 281
Grey Collie Syndrome, 86
Greyhound Review, 99
gynandromorphs, 36
Hahn, Joan, 74
haploid, 18, 19, 22, 42, 138, 299
heifer, 34, 36
hemizygous, 42
hemophilia, 103
hermaphrodite, 36
heterosis, 143, 144, 145, 146,
  147, 279, 299
heterozygous, 21, 33, 37, 50, 58,
  59, 63, 64, 73, 78, 86, 116,
  125, 145, 257, 278, 284, 285,
  299
Hoagland, Mahlon B., iii, 155,
  301
Hoffman, Eric, 4, 5, 17, 79, 81,

249, 275, 301
homozygous, 21, 37, 43, 53, 58, 63, 64, 67, 77, 79, 86, 90, 93, 105, 106, 107, 115, 116, 125, 181, 205, 212, 213, 214, 217, 239, 240, 246, 255, 282, 285
horse, xii, xiv, 3, 7, 17, 67, 132, 174, 200, 201, 205, 218, 219, 229, 232, 239, 240, 244, 259, 265, 282, 283
Huacaya, 5, 8, 72, 73, 78, 79, 80, 106, 259, 275, 276, 278, 279, 280, 281, 282, 284, 285, 286, 288, 289, 313
hybrid vigor, 131, 138, 140, 143, 146, 275, 276, 278
hypostatic, 205, 212, 213, 255
inbreeding, 103, 104, 105, 106, 107, 108, 109, 110, 113, 116, 118, 121, 122, 125, 127, 143, 181, 184, 277, 291, 299
inbreeding depression, 108, 127, 143
incomplete dominance, 67, 68, 283
incomplete penetrance, 71, 73, 74, 263, 284
independent assortment, 22, 140
Irish Wolfhounds, 136
Jacob sheep, 173, 219
Kadwell, Miranda, 5
Kendall, Grant, 139, 182
Kerr, W. MacKintosh, 13, 161, 189, 220, 231, 245
King, Robert C., 3, 6, 97, 260
Koenig, Julie, 17, 85, 209, 224, 225, 228, 229, 240, 242
Lama Letter, 169, 174
lethal, 63, 64, 67, 87, 151, 152, 161, 239, 250
Levine, Joseph, 34, 35, 162

linebreeding, 93, 104, 105, 107, 108, 110, 113, 116, 117, 118, 121, 123, 140, 143, 277, 279, 291
linecrossing, 3, 127, 128, 140, 145, 277
Little, Clarence C., 11, 12, 162, 209, 211, 212, 255, 257
llamas, vi, xi, xiii, 3, 4, 5, 12, 25, 85, 94, 98, 109, 117, 155, 162, 177, 189, 198, 201, 202, 206, 220, 222, 223, 224, 225, 226, 227, 228, 230, 234, 236, 238, 240, 244, 245, 264, 280, 286
locus, 21, 23, 28, 53, 60, 63, 67, 71, 80, 83, 84, 107, 116, 139, 157, 200, 202, 204, 205, 209, 210, 211, 212, 213, 214, 215, 216, 217, 218, 219, 220, 224, 226, 227, 228, 229, 230, 231, 232, 233, 237, 238, 240, 243, 244, 245, 246, 253, 254, 255, 256, 257, 259, 260, 261, 282, 283, 289, 299
Lowe, Bruce, 90, 91
Lysenko, Trofim, 89
malaria, 150, 164
McClintock, Barbara, ix
meiosis, 19, 23, 138
melanin, 87, 197, 201, 216, 218, 259, 260, 262, 265
Mendel, Gregor, ix, x, xii, 15, 16, 26, 36, 84, 236, 237, 293
Mendelian, 28, 53, 67, 71, 74, 83, 84, 85, 157, 162, 181, 285
Merino, 131, 172
merle, 215, 240
micron count, 84, 85, 172, 194
mitochondrial DNA, 16, 19, 33, 37, 94, 266
mitosis, 18

Morgan horse, 183
multiple allelism, 21, 22, 209, 210, 225, 243, 246, 255
mutations, vi, 11, 27, 35, 151, 161, 162, 163, 164, 165
neoteny, 175
nucleotides, xiii, 25, 27, 28, 29, 155
nutrition, 85, 151, 190, 194, 199, 222
outbreeding, 127, 128, 140, 145, 146, 279, 291
outcrossing, 145, 146, 277, 278
ovaries, 20, 34
Patino, Maria, 35
pedigree, 23, 74, 90, 91, 92, 93, 99, 108, 109, 114, 116, 121, 122, 123, 124, 125, 127, 128, 135, 138, 140, 181, 182, 183, 184, 185, 186, 253, 255
phaeomelanin, 201, 229
phenotype, 6, 7, 8, 21, 22, 49, 50, 51, 57, 58, 67, 71, 73, 79, 80, 97, 98, 110, 116, 121, 129, 132, 138, 140, 161, 168, 177, 186, 200, 201, 206, 216, 218, 222, 228, 236, 238, 245, 259, 265, 271, 277, 278, 282, 283, 299, 300
piebald, 199, 203, 217, 223, 240, 241, 242, 243, 244, 245, 246, 253, 255
pigment, 7, 60, 67, 83, 86, 162, 201, 202, 203, 204, 211, 212, 213, 216, 218, 219, 226, 227, 229, 232, 234, 237, 238, 245, 246, 248, 249, 250, 254, 255, 259, 260, 261, 263, 265, 267, 271, 282, 283
Platt, Barbara, 172
pleiotropic, 41, 87

Pollack, Robert, ix
polydactyly, 74
polygenes, 41, 83, 85, 139, 152, 213, 259, 264, 284
polygenic, 74, 81, 83, 85, 97, 144, 200, 284
Postlethwait, John H., 30, 303
prepotent, 90, 113, 115, 183, 277
primordial, 19, 20
Principle of Segregation, 22
probability, 57, 58, 73, 78, 79, 80, 106, 150, 246, 247
progeny, 54, 58, 63, 73, 89, 90, 105, 106, 113, 115, 116, 143, 144, 158, 183, 234, 288
protein, 16, 17, 26, 27, 28, 29, 155, 163, 194
Punnett Square, 51, 52, 64, 73, 78, 79, 105, 116, 158, 230, 232, 247, 254, 285, 300
qualitative, 85
quantitative, 83, 84
Queen Victoria, 103
recessive, 6, 7, 12, 29, 42, 50, 51, 52, 53, 57, 58, 59, 60, 63, 74, 77, 78, 79, 80, 81, 83, 85, 86, 103, 106, 109, 145, 149, 151, 162, 168, 198, 205, 210, 211, 212, 213, 214, 216, 217, 218, 220, 224, 225, 226, 227, 229, 230, 231, 233, 234, 235, 239, 240, 243, 244, 245, 246, 247, 255, 256, 257, 259, 261, 263, 277, 278, 282, 283, 284, 285, 300
RNA, vi, 24, 25, 29, 155, 300
roan, 198, 225, 230, 232, 233, 234, 236, 237, 238, 239, 255
ruminants, 162, 163, 219
Saluki, 3, 7, 138
serotonin, 204

sex chromosomes, 20, 33, 35, 36, 51
sex-limited, 37, 38
sex-linked, 36, 37, 43, 103, 227, 266
sheep, vi, xiii, 4, 6, 17, 19, 22, 27, 84, 94, 98, 109, 131, 133, 140, 145, 167, 168, 172, 173, 174, 177, 178, 181, 190, 201, 206, 219, 224, 228, 240, 244, 264, 279, 280, 303, 313
Sickle Cell Anemia, 149, 150, 164
Simmons, Puala, 131, 172, 177, 303
skewbald, 243, 244
somatic cells, 18, 19, 20, 29, 38, 43, 164
species, xii, xiii, 3, 4, 5, 6, 7, 16, 17, 22, 29, 30, 33, 34, 36, 41, 44, 45, 49, 51, 53, 64, 74, 84, 86, 94, 98, 100, 101, 106, 109, 125, 129, 132, 133, 135, 143, 144, 146, 151, 152, 153, 155, 156, 157, 158, 164, 168, 169, 171, 174, 175, 178, 181, 183, 185, 193, 198, 200, 201, 203, 205, 206, 209, 210, 211, 216, 217, 218, 219, 220, 226, 227, 229, 230, 236, 237, 238, 239, 240, 242, 244, 246, 249, 250, 259, 260, 261, 262, 263, 264, 270, 272, 275, 276, 282, 291, 299
sperm, 16, 17, 18, 19, 20, 22, 33, 34, 36, 38, 94, 138, 143, 299
sperm cells, 16, 17, 19, 22, 33, 34, 299
Spiderleg syndrome, xiii
Sponenberg, D. Phillip, v, 3, 6, 13, 79, 81, 113, 131, 133, 162, 171, 174, 200, 202, 205, 209, 222, 224, 225, 226, 227, 228, 230, 231, 233, 237, 238, 239, 240, 241, 249, 250, 257, 259, 263, 283, 288, 289, 301
Srb, 13, 22, 106, 107, 125
SRY, 35
Strain, George M., 261
Suri, 5, 72, 73, 78, 79, 80, 81, 106, 133, 275, 276, 278, 279, 280, 281, 282, 283, 284, 285, 286, 287, 288, 289
Suzuki, David, 34, 35, 162
thyroid, 193
ticking, 22, 216, 255
Tilghman, Shirley M., v, 34, 41, 42, 43, 44
transgenic, 6
tuxedo, 197, 199, 232, 233, 238, 240, 241, 242, 243, 244, 246
tyrosinase, 260
utero, 36, 63, 151, 163
Vicuñas, 176
Whippet, 7, 50, 59, 90, 100, 104, 113, 163, 191, 197, 199, 201, 214, 253, 256, 269
wool blind, 172, 174
Zygote, 300

About the Authors

Ingrid Wood was born and raised in Germany. In 1970, she followed her American-born husband to the USA. After raising their son Benjamin in the quaintly historical town of Mount Holly, the Woods currently live happily at Stormwind Farm in Springfield, New Jersey. There they tend to a slowly expanding herd of Huacaya alpacas and their beloved Whippets. Wood's interests include writing, reading non-fiction, spinning the fine fiber of her alpacas, gardening, and lure coursing as well as racing the Stormwind Whippets. Her articles have been published in *The LAMA LETTER*, *The GALA Newsletter*, *American Livestock Magazine*, *MAPACA Newsletter*, *Alpacas Magazine*, and the German camelid publication *LAMAS*. Wood's fascination with livestock breeding traces back to wonderful childhood vacations spent with relatives who kept bees, milked their dairy cows, and raised pigs, chickens, and small flocks of sheep.

Denise Como is a life-long resident of New Jersey. She became interested in purebred dogs as a teenager in 1966, when she purchased her first Collie. The diverse group of sighthound breeds captivated her attention shortly thereafter. Having owned and bred Borzoi for decades under her Wolfwind prefix, Denise and her husband also share the house with Whippets, Salukis of desert lineage, occasionally Greyhounds, and the remains of a small colony of rare sighthounds from India known as Rampur Hounds. Their two sons and their daughter are grown with families of their own.

Como's first sighthound handbook for beginners, *So You Want to Run Your Sighthound*, was published in 1996. It is in its sixth revision (*Running Your Sighthound*, 2003). Her second book, *Sighthounds Afield — The Complete Guide to Sighthound Breeds and Amateur Performance Events*, includes a comprehensive chapter on adopted ex-racing Greyhounds (2003, 1st Books Library, Bloomington, Indiana). Como has been a contributing writer for a number dog magazines, including the *AKC Gazette*, *AKC Afield*, *AKC Courser!*, *Celebrating Greyhounds*, *A Breed Apart*, *Chart Polski Heartbeat*, *To The Line*, *The IG Times*, and others. She is a fully licensed ASFA and AKC lure coursing judge and enjoys judging at amateur race meets.